# 30-SECOND
# METEOROLOGY

# 30-SECOND METEOROLOGY

The 50 most significant events
and phenomena, each explained
in half a minute

Editor
**Adam A. Scaife**

Foreword
**Julia Slingo DBE FRS**

Contributors
**Edward Carroll**
**Leon Clifford**
**Chris K. Folland**
**Joanna D. Haigh**
**Brian Hoskins**
**Jeff Knight**
**Adam A. Scaife**
**Geoffrey K. Vallis**

Illustrations
**Nicky Ackland-Snow**

**METRO BOOKS**
New York

**METRO BOOKS**
New York

An Imprint of Sterling Publishing
1166 Avenue of the Americas
New York, NY 10036

This book was conceived,
designed, and produced by
**Ivy Press**
210 High Street, Lewes,
East Sussex BN7 2NS, U.K.
www.ivypress.co.uk

Publisher  **Susan Kelly**
Creative Director  **Michael Whitehead**
Editorial Director  **Tom Kitch**
Art Director  **Wayne Blades**
Commissioning Editor  **Stephanie Evans**
Senior Project Editor  **Caroline Earle**
Designer  **Ginny Zeal**
PIcture Researcher  **Katie Greenwood**
Glossaries Text  **Leon Clifford**

ISBN 978-1-4351-6343-0

For information about custom
editions, special sales, and premium
and corporate purchases, please contact
Sterling Special Sales at 800-805-5489
or specialsales@sterlingpublishing.com.

Typeset in Section

Manufactured in China

10  9  8  7  6  5  4  3  2  1

www.sterlingpublishing.com

# CONTENTS

# FOREWORD
## Professor Dame Julia Slingo DBE FRS

Our planet's atmosphere is massively complex, and as a result the weather we experience varies hugely from place to place and over different times of the year. From heatwaves to storms to blizzards, weather and climate affect how we live and everything we do.

Through human ingenuity we have adapted to live with the weather: growing crops that will flourish, building homes that withstand local conditions, and planning our lives around the seasons. However, throughout history extreme weather events, such as droughts, floods, and extreme cold, have challenged societies' resilience, costing lives and livelihoods.

So it is natural that we have sought to understand our weather and climate—what causes it to fluctuate and change over hours, weeks, seasons, and years. And this endeavor has led to increasingly skillful weather forecasts and climate predictions that allow us to prepare for what's coming—whether it is heavy, thundery rain this afternoon, a damaging wind storm later this week, the chance of a colder winter this year, or more extreme heatwaves in the coming years as our climate warms.

Today, we live in a global economy, relying on global trade, efficient transport networks, and resilient and reliable provision of food, energy, and water. All of these systems are vulnerable to adverse weather and climate. The additional pressure of climate change creates a new set of circumstances and poses fresh challenges about how secure we will be in the future. More than ever, the weather and climate have considerable direct and indirect impacts on us—our livelihoods, property, health, well-being, and prosperity.

Through the application of scientific rigor and the use of cutting-edge technologies, such as satellites and supercomputers, the study of meteorology has revolutionized our understanding of the weather and climate we experience and enabled us to forecast its future behavior with ever-increasing skill. From the global to local and from hours to decades, our understanding of weather and climate and the predictions we make enable us plan for the future. Dip into these pages to learn more about the science of meteorology and how our weather works.

**Tackling the greenhouse** *effect will test human ingenuity, making scientific knowledge and understanding of our planet and its atmosphere even more pressing during this century.*

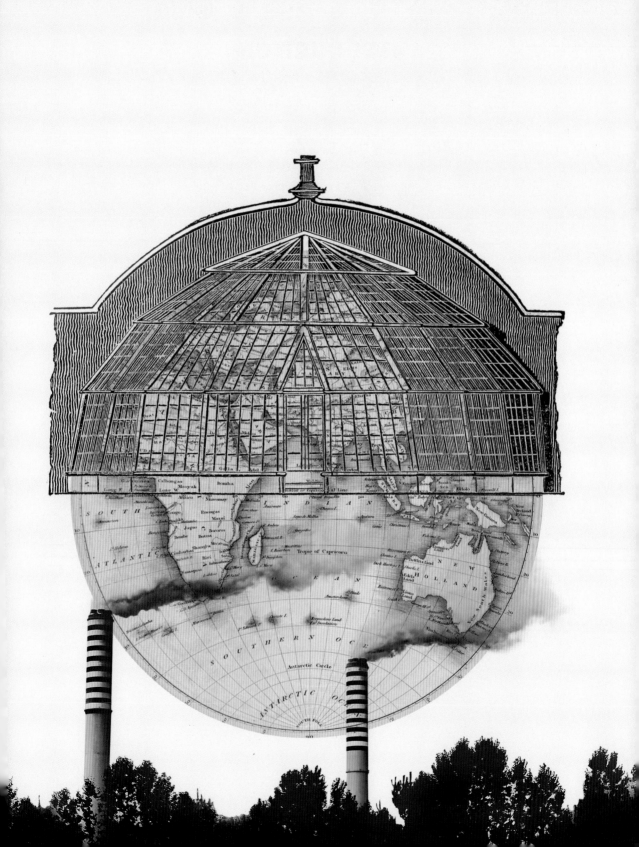

# INTRODUCTION
## Professor Adam A. Scaife

The atmosphere is inextricably linked with human
endeavor. There are countless examples: the early loss of ships at sea
in fierce storms; fortuitous Trade Winds that aided the crossing of vast
oceans; devastating droughts, floods, and hurricanes that still take
thousands of lives even today; and the reliance on the regular onset of
rainy seasons that sets the timing of agriculture—the weather has an
enormous impact on us all. It has shaped cultures and is even responsible
for key turning points in history, such as the failure of Napoleon's march
into Russia in the grip of winter in the nineteenth century or the mass
displacement of farmers in America due to the Dust Bowl drought
of the 1930s.

These events and their consequences come about because the
atmosphere is constantly changing. It varies on all timescales, from the
afternoon sunshine that brings us outdoors to the prolonged rains of
the tropical monsoons. There are even decades where the weather keeps
reverting back to similar behavior summer after summer or winter after
winter, only to change and be followed by years of the opposite. This
apparently mysterious behavior arises because the atmosphere is actually
a constantly circulating fluid; trapped in a thin layer at the surface of our
rotating Earth it swirls and flows just like the water in your bathtub.
Apart from a lack of internal friction, the equations that govern the
future behavior of our weather and climate are almost the same as the
equations that apply to the swirling fluid in your morning cup of coffee.

There are just six mathematical equations that encapsulate all of
this and they can be written so concisely that they fit onto the back of
a postcard. In fact, given their enormous significance, it is surprising
they are not printed on T-shirts! These few equations that determine
the fate of our weather also stem from well-established, or even old
physics: they derive from Newton's laws of motion, the physics of heat
and gases that was uncovered more than a century ago, and the fact that
air is neither created nor destroyed as it blows around with the wind.

Yet despite the fact that we know so well the equations that govern what the future weather will be, they are still shrouded in uncertainty. This is because the problem is not like your school math exercise where pencil and paper will do. Instead the equations are intractable: they simply cannot be solved precisely. Even worse, they show that small changes can eventually give rise to big impacts through "chaos" and the revelation that a seagull flapping its wings can actually result in a hurricane months later fundamentally puts limits on any forecast. This is what makes meteorology so challenging and yet so compelling, and understanding the fluid that is our atmosphere is one of the most active remaining areas of terrestrial physics.

Modern advances in understanding and the production of improved forecasts from hours to years ahead are now completely dependent on modern technology. A suite of scientific instruments measures various

**Harnessing the power** *of the wind, constantly circulating the planet, is one early example of how the human race propelled itself toward becoming a global society.*

parameters of the Earth's atmosphere. Arrays of environmental satellites continuously monitor the behavior of the atmosphere and oceans and relay their measurements back to Earth: polar orbiters constantly circle the Earth in an almost north–south orbit at 500 miles altitude and geostationary satellites are sited over 22,000 miles above the surface.

These vital observations, along with terrestrial measurements from weather stations, radar, aircraft, and weather balloons are automatically incorporated into vast databases that constantly update our changing picture of the global weather. Some of the most powerful computers on the planet are used to pick up this latest information, combine it with computerized representations of the underlying equations, and calculate what will happen next. The results are remarkable: the computer models produce virtual simulations of the Earth's global weather that contain just about everything we see, from jet streams and hurricanes to multi-year climate oscillations between the ocean and atmosphere like El Niño. All of these features emerge spontaneously from the few fundamental equations at the core of the computer models. The results of this daily application of basic science drive everything from the forecast made this evening by your local TV weather forecaster to the century-long projections that affect government policies on climate change.

This book takes you to the edge of our current knowledge. It is written by leading experts in weather forecasting, the physics of the atmosphere, and the behavior of global climate, who bring you their insights after more than 200 years of combined research experience. In the first section, **The Elements**, the groundwork is done to describe the basic features of our weather with detailed explanations of how many common meteorological phenomena that we all take for granted are actually produced. **The Global Atmosphere** follows, providing the big picture in which our weather sits, identifying everything from the jet streams that deliver storms across the ocean basins to the return flow in the tropical Trade Winds. The third section, **The Sun**, explains the science behind the ephemeral optical phenomena that light up our skies and the way our local star determines and profoundly influences the weather and climate on Earth. **Weather Watching & Forecasting** gives an insight into the machinery that is brought into action every day to make your weather forecast. Section four looks further ahead and asks the question **Can We Change the Weather?** using examples from historical changes in the ozone layer to future climate change. All of this is put in the context of natural weather fluctuations in **Weather Cycles** before the final section, **Extreme Weather**, explains the wilder side of meteorology.

You can open this book at any point or immerse yourself in a complete chapter at a time and also read the fascinating histories of some of the great pioneers in meteorology. However you approach the 50 topics, I urge you to speculate on what is to come. Before the advance of atmospheric science, weather forecasters were ridiculed for their dream of predicting the future, but the ever-increasing accuracy of weather forecasts today is a reality and this makes them an essential tool to society worldwide. Even long-range predictions of the average weather from months to years ahead are now possible and, in some cases, point toward dramatic future events. Some of these forecasts are destined to become ever more critical and they are unfolding right now, as you read this book, when the globe is warmer than ever recorded before.

**Basic physics encapsulated**
*in a few fundamental equations contain the secrets of our future weather and climate, from winter cold snaps to summer heatwaves.*

# THE ELEMENTS

**aerosols** An aerosol consists of billions of minute droplets of liquid or particles of solid suspended in a gas. Pollen, sea salt, and soot from combustion can all make aerosols in the atmosphere. Pollution and volcanic eruptions can lead to aerosols high in the stratosphere that consist of tiny droplets of sulfuric acid. Some aerosols (e.g. soot) absorb incoming solar radiation and so act to warm the Earth; others (e.g. sulfuric acid droplets) can reflect solar radiation back into space and so have a cooling effect. Aerosols can act as the seeds around which cloud droplets form. The size of aerosol particles can range from 1 nanometer (one billionth of a meter) to 100 micrometers (one ten-thousandth of a meter).

**equinox/equinoctial seasons** The equinoctial seasons contain the equinoxes—when day and night are equally long—which occurs twice yearly, on or around March 21 and September 21. They are the seasons between winter and summer; so spring and fall are equinoctial seasons. In meteorological terms equinoctial seasons are usually considered to be the three-month periods of March, April, and May and September, October, and November. The equinoctial seasons are transition seasons between the more extreme summer and winter.

**longwave radiation** The heat radiated by the warm surface of the Earth and warm areas in the atmosphere. It has a longer wavelength than the visible and ultraviolet light from the Sun which is known as shortwave radiation. Longwave radiation is infrared and invisible but it is a form of electromagnetic radiation just like light and radiowaves.

**pressure gradients** The change of atmospheric pressure with distance in a given direction. This gradient results in a force that acts on the air in a direction that is at right angles to the isobars—the lines of equal air pressure seen on weather maps. This force tries to push the air from areas of high atmospheric pressure toward areas of lower pressures and it is a source of wind. The steeper the pressure gradient, the more tightly packed the isobars, and the stronger the resulting wind. In meteorology, the idea of pressure gradients is applied to the behavior of the atmosphere and is usually measured in millibars per kilometer (mb/km)—a millibar being a unit of atmospheric pressure. The nominal atmospheric pressure of the Earth at sea level is 1,000 millibars or 1 bar.

**saturation** The state of the atmosphere in which the air contains the maximum amount of water vapor that it can hold at that particular temperature and air pressure.

At saturation, relative humidity—the amount of water vapor in the air compared to the amount that the air could hold—is 100 percent and further evaporation of water vapor into the air cannot occur. The capacity of the air to hold water vapor grows with increasing temperature and declines with decreasing temperature. This is why warmer climates experience greater humidity and why warm humid air forms clouds as the air rises and cools.

**solstice**  An astronomical event that occurs twice yearly, around June 21 and December 21, due to the tilt of the axis of the Earth's rotation to the plane of its orbit around the Sun. A solstice occurs in the summer and winter. In the northern hemisphere the summer solstice occurs in June and the winter solstice in December, and vice versa in the southern hemisphere. At solstice, the amount of daylight is at an annual maximum in one hemisphere and at an annual minimum in the other.

**supercooling/supercooled water**
Supercooling occurs when a liquid is cooled below its normal freezing point but does not solidify. Supercooled water droplets are found in high-altitude clouds where the temperature of the air is below the freezing point of water. This supercooled state can

only be achieved in droplets that do not contain impurities or aerosols that would otherwise act as seeds to trigger crystallization. Research suggests that the phenomenon of supercooling may be due to the molecules of water arranging themselves in a way that is incompatible with crystallization.

**temperature inversion**  In the troposphere (the lowest layer of the Earth's atmosphere) air temperature usually falls with increasing altitude but sometimes it can increase, resulting in a blanket of warm air sitting above a layer of cooler air. This is known as a temperature inversion. Rain falling through a temperature inversion can freeze, causing freezing rain. If the air below the inversion is sufficiently humid fog can form. Over populated areas, temperature inversions can act as a lid that traps pollution near the ground.

**vortex/vortices**  In meteorology, a vortex refers to a rotating mass of air, often circulating around a low-pressure system. A hurricane or typhoon is an example of a vortex of air circulating around a center of low pressure. Larger and more persistent atmospheric vortices are found circulating around low-pressure regions over each pole—and the so-called Polar Vortex over the North Pole has been famously associated with severe winters in North America and Eurasia.

# AIR

## the 30-second meteorology

### The Earth is thousands of miles

in diameter, whereas the air we breathe sits in a thin skin just 62 miles thick: a distance you could drive in under an hour. Air is a mixture of different gases, primarily nitrogen (78%) and oxygen (21%). The last 1% is inert argon, carbon dioxide ($CO_2$), and tiny amounts of other gases like ozone. There is also water vapor, around 1% at the surface but this depends where you are. The restless weather systems in the troposphere constantly stir these gases with a dash of pollutants and other chemicals, so that most of the air is well-mixed. It takes under a year for mixing to occur globally, which is why increasing carbon dioxide from cities and industrial centers can be measured almost anywhere. Although a tiny fraction of the air, $CO_2$ affects the Earth's temperature and its concentration is increasing rapidly through human activities, driving global warming. These changes are happening fast but the balance of gases was not always the same. The distant geological past contains very long periods with far less oxygen and other periods with more, which had remarkable effects, including enabling insects to grow to many times their present size—one example of how the makeup of the air is intimately linked with life on Earth.

**3-SECOND BREEZE**
The air is a thin layer of different gases constantly stirred by the weather, which mixes it up on a timescale of months.

**3-MINUTE SHOWER**
Air is gradually circulating between the lower atmosphere and the stratosphere where the ozone layer sits. It rises in the tropics and sinks over the poles but this process is very slow and it takes years for air molecules to complete the circuit. This slow circulation is important as it sets the time for cleansing the air of ozone-depleting chemicals.

**RELATED TOPICS**
See also
LAYERS OF THE ATMOSPHERE
page 18

THE OZONE HOLE
page 108

GLOBAL WARMING &
THE GREENHOUSE EFFECT
page 110

**3-SECOND BIOGRAPHY**
JOHN TYNDALL
1820–93
Irish physicist who discovered numerous physical properties of the air including how molecules interact with infrared heat radiation to warm and cool the atmosphere

**30-SECOND TEXT**
Adam A. Scaife

*The composition of air is vital in ensuring there is life on Earth, blocking the deadly rays of the Sun, trapping heat to keep the environment comfortable, and, crucially, providing the oxygen we breathe.*

# LAYERS OF
# THE ATMOSPHERE

## the 30-second meteorology

## RELATED TOPICS
See also
AIR
page 16

THE OZONE HOLE
page 108

## 3-SECOND BIOGRAPHY
ARISTOTLE
384–322 BCE
Greek polymath who wrote the
first book about meteorology
in ca. 350 BCE in which he
outlined the hydrological cycle
and discussed numerous
weather phenomena

## 30-SECOND TEXT
Adam A. Scaife

## 3-SECOND BREEZE
All of our weather occurs
in the lowest layer of
the atmosphere—the
troposphere—above
which sits the quiescent
stratosphere and the
tenuous mesosphere.

## 3-MINUTE SHOWER
Other planets also have
a troposphere and a
stratosphere and the
boundary between them
often occurs at about
the same pressure as in
the Earth's atmosphere.
Jupiter is one example but
the balance of heating is
different in the atmosphere
of this gas giant, because
much of its energy comes
from a mysterious heat
source beyond observation,
deep in the Jovian
atmosphere.

Let's take a journey up through
the atmosphere. If you have ever climbed a
mountain you will know that the air gets colder
the higher you go, by about 3°F every 1,000
feet. This is because sunlight is absorbed
by the ground, making it warmer there. Energy
spreads upward by reradiating heat and by warm
moist air rising up through the turbulent
overturning troposphere. The rising air causes
much of our weather, with clouds regularly
reaching 9 miles in the tropics. But what
happens if we go higher? The air can't just keep
getting colder.  As we ascend farther and the air
thins to just a tenth of its density at the surface,
it starts to warm again; we have entered the
stratosphere. Here the ozone layer warms the
atmosphere by absorbing ultraviolet light from
the Sun. As warmer air is now sitting atop cooler
air, everything is stable—there is no weather
here. Eventually the ozone starts to thin and
temperatures begin to drop again: we are in the
mesosphere. Here the air is 10,000 times thinner
than at the surface and it is turbulent again.
Ripples travel up from distant weather systems
below and drive winds that pull air upward in
summer to create the coldest point in the
atmosphere, more than –150°F.

*Our atmosphere has
four layers, based on
temperature. The layer
our weather is confined
to—the troposphere,
closest to Earth—is
the warmest. Highest
lies the thermosphere,
the realm of meteors
and auroras.*

# SEASONS

## the 30-second meteorology

### The axis upon which the Earth

spins each day is tilted at 23.4 degrees to the Earth's orbit around the Sun. It has a fixed direction in space, so as the Earth makes its annual journey around the Sun, the northern hemisphere is tilted toward the Sun for half of the year, and the southern hemisphere is tilted toward the Sun for the other half. This causes the amount of daylight to increase and decrease, and makes the Sun rise higher or lower in the sky, changing its ability to warm the surface. Outside of the tropics, this creates a cycle of temperatures associated with four seasons—spring, summer, fall, and winter. In the tropics, the midday Sun is always high in the sky, and temperature varies little through the year. Here, the seasons are defined by changes in rainfall rather than temperature. The tropical rainfall zone tracks the shifting latitude where the Sun is overhead, which typically results in a short wet season and longer dry season, as in many parts of India. However, some locations, such as East Africa, experience two wet seasons corresponding to the northward and southward passage of the overhead Sun.

**3-SECOND BREEZE**
It is the tilt of the Earth's axis that produces the seasons rather than the distance of the Earth from the Sun.

**3-MINUTE SHOWER**
Astronomers can precisely measure the seasons by the solstices and equinoxes, which are key moments in the Earth's annual journey around the Sun. But in terms of the weather, the seasons tend to change more gradually. Meteorologists' seasons are therefore always sequences of calendar months, which are the building blocks of climate statistics. The sets of months are chosen for the region in question, such as the four three-month seasons used in mid-latitudes.

**RELATED TOPICS**
See also
RAINY SEASONS
page 64

MONSOONS
page 66

MILANKOVITCH CYCLES
page 138

**30-SECOND TEXT**
Jeff Knight

*Antonio Vivaldi would not have been inspired to compose his* Four Seasons *had he lived near the equator. The tilt of the Earth's axis explains why seasons are more apparent in the mid-latitudes in contrast to the tropics where seasonal change is less marked.*

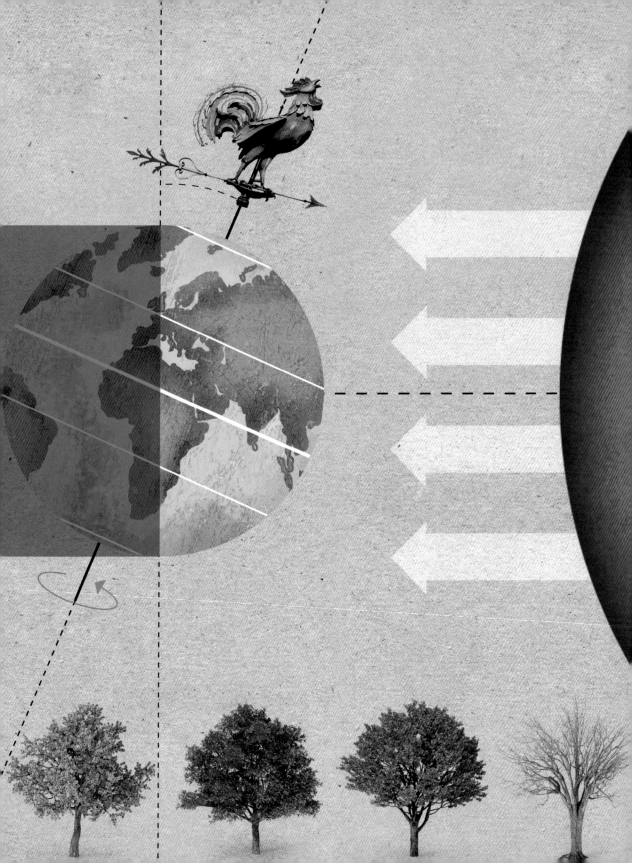

# CLOUDS

## the 30-second meteorology

**3-SECOND BREEZE**
Clouds are composed of tiny water droplets or ice particles, each of which forms around an aerosol, a minuscule, solid particle.

**3-MINUTE SHOWER**
In 1802 Luke Howard, an amateur meteorologist, turned a lifelong interest in staring out of the window at the sky to good use and published a cloud classification, *On the modification of clouds*. Cloud types today are named after the terms he proposed, such as *cumulus*, for vertically extensive clouds, *cirrus* for wispy clouds, *stratus* for layered clouds, and *nimbus* for rain clouds.

Water vapor is a gas, invisible but present nearly everywhere in the atmosphere in varying concentrations. Cooling air reduces its capacity to contain water vapor and, if it is chilled sufficiently, saturation occurs. At this point water starts changing from its gaseous to its liquid or ice phases. The commonest cooling mechanism is lifting, for instance as warm air rises above a wedge of cold air at a front, or when bubbles of air rise over ground heated by the Sun. Decreasing pressure causes a rising parcel of air to expand, the work done using up heat energy on much the same principle that a fridge uses. Liquid water produced by chilling collects in minute droplets on surfaces, apparent as misting on a glass of iced water. In the atmosphere, the surfaces required for condensation are provided by specks of material known as aerosols. They have many sources, including salt particles released by breaking ocean waves, and industrial pollution. All cloud droplets contain microscopic nuclei of this sort and grow by condensation (or deposition in the case of ice) to a size of the order 1–10 microns (millionths of a meter). Being tiny, they have negligible fall-speeds, and remain in effect suspended, despite bulk cloud weights of millions of tons.

**RELATED TOPICS**
See also
RAIN
page 24

SNOW
page 28

HAIL
page 30

FOG
page 32

**3-SECOND BIOGRAPHY**
LUKE HOWARD
1772–1864
British pharmacist and amateur meteorologist who proposed a nomenclature of clouds in 1802

**30-SECOND TEXT**
Edward Carroll

*Luke Howard's original cloud classification was further refined by combining types and including reference to three height forms, giving rise to such denominations as altocumulus, cirrostratus, and cumulonimbus.*

# RAIN

## the 30-second meteorology

To become a raindrop, a cloud droplet has to increase in mass about a million fold. Droplets of differing sizes gradually settle out at different speeds and can grow by colliding and coalescing. This is generally a slow process, but a small proportion of droplets have a high enough number of chance collisions to grow and acquire significant fall-speeds, allowing them to sweep up an increasing number of smaller droplets—an accelerating process which can produce a raindrop within 20 minutes. Outside the tropics, another process dominates. Here, most rainbearing clouds are below 32°F, but unless very cold (below –4°F), just a few of them will freeze. Instead, the bulk of the rain cloud consists of droplets of supercooled water below 32°F. Deposition of water vapor onto ice takes place more readily than condensation into water, so the few ice particles grow quickly, drawing water vapor from the air and causing the many supercooled water droplets to shrink by evaporation. Rapidly growing in size at the expense of cloud water, ice particles fall, collecting supercooled water droplets on their way, which freeze onto them. As they descend to warmer levels they melt, and all of this is forgotten as they reach the ground as rain.

### 3-SECOND BREEZE
Most extra-tropical raindrops start life as ice particles in the cold, upper reaches of clouds, sweeping out cloud droplets as they fall and melt.

### 3-MINUTE SHOWER
In diameter, typical cloud droplets are about 1–10 microns (millionths of a meter), drizzle drops 100–500 microns, and raindrops 500–5,000 microns. Small cloud droplets, being closer in size to the wavelength of visible light, scatter it back more efficiently than large cloud droplets or raindrops, which absorb more of the light. Clouds therefore often look darker and more menacing as droplets approach raindrop-size.

### RELATED TOPICS
See also
CLOUDS
page 22

MONSOONS
page 66

RAINBOWS
page 78

ACID RAIN & ATMOSPHERIC POLLUTION
page 114

### 30-SECOND TEXT
Edward Carroll

*Without clouds there is no rain but the droplets that form clouds are too small to fall—until they coalesce with other droplets and become heavier. When those drops are $1/64$ inch in diameter or larger, the heavens open, making us grateful for a roof over our heads.*

# FROST

## the 30-second meteorology

Climatologically, a frost is the occurrence of a temperature below 32°F, the melting point of ice. Temperature can vary strongly with height, and standard recording practice is to measure it around 5 feet above ground level. Night-time cooling of the ground by longwave radiation on clear, calm nights results in an inversion of the usual tendency for temperature to fall with height; in these circumstances ground temperature can be 9°F below that at 5 feet. So ground frost forms more easily than air frost, especially over grassy surfaces, where air trapped between grass blades provides insulation from heat stored in the ground. A late spring frost can nip tender, low-growing plants in the bud because the freezing of water causes expansion, rupturing cell walls. Hoar frost is a visible manifestation of frost, being ice crystals deposited directly from water vapor onto subzero vegetation and other surfaces. With high humidity and a breeze, fresh supplies of water vapor are constantly brought into contact with cold surfaces and a thick coating of hoar frost can accumulate, occasionally producing a magical, wintry landscape. Clearer ice results when dew freezes, or when supercooled fog droplets spontaneously freeze on contact with surfaces, in the latter case known as rime.

## RELATED TOPICS

See also
CLOUDS
page 22

SNOW
page 28

FOG
page 32

**30-SECOND TEXT**
Edward Carroll

**3-SECOND BREEZE**
When the temperature falls below 32°F, a frost is said to occur, and if conditions are right, it is marked by the formation of ice.

**3-MINUTE SHOWER**
From the sixteenth to the nineteenth century, a period known as the Little Ice Age, winters in London, England, were sometimes severe enough to freeze the Thames River. The opening up of this wide thoroughfare through the city stirred excitement in the populace and resulted in spontaneous "frost fairs" during which stalls and entertainments were set up, oxen were roasted, and even, on one occasion, an elephant was led across the river.

*The crystalline structure of ice manifests itself in many different guises, from delicate, lacy, and ephemeral forms to a lattice of immense rigidity—strong enough to bear the weight of man and beast.*

# SNOW

## the 30-second meteorology

Cloud ice particles form on aerosols known as freezing nuclei, which have a similar, hexagonal, small-scale structure to ice crystals. Deposition from vapor into the ice phase can result in the growth of astoundingly complex, geometric, even organic-looking structures—basically hexagonal, but needlelike, platelike, or intricate, fernlike branching shapes, depending on temperature and humidity. The branching forms easily interlock and aggregate as they fall, reaching the ground as classic, feathery-looking snowflakes if the temperature is close to or below 32°F. Falling ice particles can also grow by gathering supercooled water droplets, which freeze on contact and stick to give a more irregular shape, masking the original crystalline form. This process is called accretion and the resulting precipitation graupel or, when consisting of small particles, snow grains. In practice, snow is often composed of a mixed type resulting from combined aggregation and accretion. Due to trapped air, snow accumulates to depths 10–15 times greater than the same mass of liquid water—for example, $^{25}/_{64}$ inches rainfall equivalent of snow lies 4–6 inches deep. A blizzard occurs when falling or lying snow is whipped up by strong winds. It swirls around, piling in hollows and against obstacles into drifts, burying vehicles and livestock.

**3-SECOND BREEZE**
Tiny, often complex, hexagonal ice crystals grow when water vapor deposits as ice onto aerosols, forming snow as they collide and interlock.

**3-MINUTE SHOWER**
Snow can be difficult to forecast in marginal winter climates such as northwest Europe's, since an error of 1.8°F in temperature can make the difference between $^{25}/_{64}$ inches of rain with little impact, and traffic chaos from 6 inches of snow. The challenge is made harder because temperature falls more with increased intensity of rain, so that what is forecast to be moderate rain can turn to disruptive snow if it is heavier than expected.

**RELATED TOPICS**
See also
CLOUDS
page 22

RAIN
page 24

HAIL
page 30

**30-SECOND TEXT**
Edward Carroll

*A farmer from Vermont, Wilson Bentley, began taking snowflake photographs in 1885 and captured more than 5,000 over his lifetime using a microscope rigged to a camera. He died of pneumonia after walking home through a blizzard.*

# HAIL

## the 30-second meteorology

If a parcel of air is raised, it expands, cools, and normally finds itself colder and denser than its environment, so sinks to its original position. However, when the atmosphere is cooled aloft, or warmed from below, it can become unstable. Its temperature falls with height faster than that of a rising parcel, which is therefore lighter than its surroundings and continues to ascend as a buoyant bubble—a process called convection. Water condenses and releases heat, further adding to buoyancy, and when ice forms the cloud becomes a cumulonimbus. Towering up to 6–9 miles tall, bulging cauliflower-like protuberances at its boundaries mark the edges of individual rising currents, and more fuzzy edges aloft signify that ice particles predominate. These hailstone "seeds" grow fast and soon start to fall, supercooled water droplets freezing onto them. If horizontal wind varies strongly with height, warm updrafts can remain separated from cold, precipitation-induced downdrafts which would otherwise eliminate them. Powerful updrafts support rapid growth of the hailstone from ice precipitation, which can recirculate as it exits the top of, then reenters, a sloping updraft. Fresh layers of ice form, sometimes alternating clear and opaque due to varying water droplet concentrations in different regions of the cloud, before the hailstone finally plummets to Earth.

**3-SECOND BREEZE**
Cumulonimbus clouds with strong, vertical currents can support lumps of ice, exceptionally over 4 inches in diameter and weighing over 2 pounds.

**3-MINUTE SHOWER**
Large hail results in significant annual crop damage in some areas. In urban locations in the developed world, the biggest financial costs occur, with over $1 billion attributed to individual storms in the U.S., Europe, and Australia. Elsewhere, for example rural India, Bangladesh and China, where people are more likely to be caught without shelter, there are records of multiple fatalities in hailstorms.

**RELATED TOPICS**
See also
CLOUDS
page 22

RAIN
page 24

SNOW
page 28

THUNDERSTORMS
& LIGHTNING
page 144

TORNADOES
page 150

**30-SECOND TEXT**
Edward Carroll

*A hailstone accumulates layer upon layer of ice as it is recirculated through the cloud, supported against falling by updrafts of between 80 and 160 feet per second.*

# FOG

## the 30-second meteorology

Fog is cloud at ground or sea level, dense enough to reduce horizontal visibility to below 3,280 feet, and frequently to less than 650 feet. When low cloud intersects high ground, hill fog results, though in general fog is unlike other clouds, which form by cooling through ascent. The principal mechanism over land is night-time emission of longwave (heat) radiation, which is most effective in the absence of cloud to radiate back. Light winds prevent warmer air mixing down from above and the resulting chilling reduces the capacity of air to contain water vapor, so water droplets condense onto tiny cloud condensation nuclei. Fog forms initially just above the ground, but deepens upward as the fog top itself becomes the radiating surface. The resulting radiation fog is commonest in valleys, where colder, denser air collects by drainage and watercourses provide moisture. Advection fog forms when moist air is chilled by flowing over a cold surface. Sea fog is a type of advection fog, commonest in spring and early summer when the sea is still cold but the air is warming up. In some dry, or seasonally dry, parts of the world, sea or hill fog is a vital source of water for trees, such as California's Coast Redwoods. Intercepted droplets coalesce on leaves or needles and drip to the ground, providing moisture to their roots.

**RELATED TOPICS**
See also
CLOUDS
page 22

RAIN
page 24

SNOW
page 28

**30-SECOND TEXT**
Edward Carroll

**3-SECOND BREEZE**
To be in fog is to enjoy a first-hand experience of having one's head in the clouds.

**3-MINUTE SHOWER**
Sooty particles from burning coal act as cloud condensation nuclei and promote fog. The combination can result in a particularly dense and persistent smoky fog—*smog*—common in nineteenth- to mid-twentieth-century London. Episodes of very poor visibility—sometimes only a few yards—known as pea-soupers were responsible for severely disrupting travel and causing respiratory problems which killed thousands.

*The concentration of water droplets in fog determines the visibility in foggy conditions. Cool ocean temperatures mean California's coasts are often subject to sea fogs which sometimes shroud the Golden Gate Bridge but also sustain forests of huge Coast Redwoods (Sequoia).*

**October 11, 1881**
Born in Newcastle upon Tyne, England

**1903**
Graduates from Cambridge with a first-class degree in the Natural Science Tripos

**1907**
Solves a problem involving water flow through peat using approximate mathematical methods applied to the differential equations of hydrodynamics

**1913**
Joins the UK Meteorological Office for the first time and investigates the use of mathematics in weather forecasting

**1916**
Conscientious objector to military service. Works in an ambulance unit in France

**1919**
Returns to work for the Meteorological Office

**1920**
Resigns from the Meteorological Office when it becomes part of the Air Ministry

**1920s**
In his spare time, conducts research into the relationship between wind and heat that produces turbulence. His equations identified what is now called the Richardson number used to predict where turbulence will occur in the atmosphere and the oceans

**1922**
Publishes a groundbreaking work, *Weather Prediction by Numerical Process*, which includes details of his pioneering hand-calculated mathematical weather forecast and his research into turbulence

**1926**
Elected to the Royal Society in recognition of his work

**1929**
Awarded a B.Sc. in Psychology from University College London

**1940**
Retires to focus on research into areas including the application of mathematics in psychology and international conflict

**1950**
Hears news of the first 24-hour numerical weather forecast made by computer

**September 30, 1953**
Dies in Kilmun, Scotland

# LEWIS FRY RICHARDSON

**Religious belief and an interest in** science shaped the life of the man who invented modern weather forecasting. Born in England into a Quaker family, Lewis Fry Richardson developed an aptitude for science which took him to Cambridge where his studies included a mix of mathematics, physics, and earth sciences—an ideal background for meteorology.

Early in his career, Richardson used the same mathematics that he would later apply to the weather in a very practical scientific problem involving the flow of water through peat. This approach, based on hydrodynamics, can calculate the evolution of constantly changing systems using the mathematics of finite differences.

Richardson was introduced to the challenges of forecasting on joining the Meteorological Office in 1913 to run the Eskdalemuir Observatory in Scotland. He recognized that weather could, in principle, be forecast using the differential equations of hydrodynamics and he set about trying to do this.

In World War I, Richardson was a conscientious objector to military service on religious grounds and left the Meteorological Office in 1916 to work in a battlefield ambulance unit. Nevertheless, he continued to develop his ideas and, in a remarkable feat of mathematics, he produced the world's first-ever numerical weather forecast by hand calculating changes in the pressure and winds at two points in central Europe over a six-hour period. Unfortunately, this forecast proved inaccurate due to the way the equations behaved.

Richardson included details of this pioneering calculation in a book published in 1922 and followed it with a series of papers establishing the theoretical basis of numerical weather forecasting. He realized that the sheer volume of calculations required for numerical weather forecasting was a major challenge. "Perhaps some day in the dim future it will be possible to advance the computations faster than the weather advances .... But that is a dream," he concluded. More graphically, describing the effect of atmospheric turbulence—another breakthrough aspect of his research—he wrote:

"Big whirls have little whirls,
That feed on their velocity;
And little whirls have lesser whirls,
And so on, to viscosity ..."

Richardson rejoined the Meteorological Office after the war but his pacifism led him to resign shortly after when, in 1920, it was amalgamated into the Air Ministry—a part of the government military establishment. In later life, he shifted his research focus to other areas as he saw the military taking a greater interest in meteorology.

Richardson lived to see the first-ever weather forecast generated by computer and, thanks to modern supercomputers, his dream of using mathematics to create weather forecasts is now an everyday reality.

*Leon Clifford*

# PRESSURE, CYCLONES & ANTICYCLONES

## the 30-second meteorology

**Although pressure comes in** many forms, in fluid dynamics and meteorology it means something quite specific: it is the force per unit area exerted by a fluid either on its container or on another part of the fluid. Which is why, for example, having too high blood pressure can cause blood vessels to burst and is thus not recommended! In the atmosphere the pressure at a point is equal, to a very good approximation, to the weight of air above that point, and so pressure decreases with altitude. Pressure varies in the horizontal too, caused by contrasts in temperature in conjunction with the Earth's rotation, leading to a never-ending pattern of pressure variations across the globe. Regions of low and high pressure are called cyclones and anticyclones, respectively. Air tends to spiral in toward cyclones and then rise, causing water vapor to condense and rain to fall. Particularly strong cyclones in low latitudes can evolve into hurricanes, with intense rain and powerful winds. In contrast, air tends to spiral outward and is replaced by sinking within an anticyclone, preventing cooling and condensation into rain and therefore giving rise to fine, clear weather.

**3-SECOND BREEZE**
Pressure is, indeed, a force of nature: it produces the winds and sometimes the rain, and if we know the pressure we (almost) know the weather.

**3-MINUTE SHOWER**
Weather is caused by the variations of global pressure patterns, for the wind and the rain can be largely inferred from these patterns: gradients of pressure produce strong winds, rain comes with cyclones, and sunshine comes with anticyclones. But these variations are chaotic—that is to say fickle and tempestuous—and the difficult science of weather forecasting lies in trying to predict what those patterns will be a day, a week, or even longer into the future.

**RELATED TOPICS**
See also
CORIOLIS FORCE
page 38

THE BALANCE OF WINDS
page 40

HURRICANES & TYPHOONS
page 146

**3-SECOND BIOGRAPHIES**
EVANGELISTA TORRICELLI
1608–47
Italian physicist who invented the barometer, essentially by measuring the weight of air by seeing how much mercury it displaced

VILHELM BJERKNES
1862–1951
Norwegian physicist, founder of the Bergen School of Meteorology, which attempted to understand and predict atmospheric motion with the aid of maps of surface pressure

**30-SECOND TEXT**
Geoffrey K. Vallis

*In the late Renaissance Evangelista Torricelli invented the barometer, allowing the study of the atmosphere to become truly scientific.*

# CORIOLIS FORCE (CF)
## the 30-second meteorology

## The Coriolis Force (CF) is an

apparent force on the air (and the ocean) due to viewing it from an Earth that is rotating. The CF is proportional to the strength of the wind and in the northern hemisphere it acts to the right of the wind. It is the CF that leads to air moving around a low-pressure system rather than being accelerated inward. In counterclockwise motion around the low, the CF to the right of the wind is outward and balances the pressure force inward. In the southern hemisphere the CF is to the left of the wind and there is clockwise motion around a low. To understand the nature of the CF, imagine a ball that is stationary on a spinning roulette wheel. If the slope of the wheel is just right, the outward centrifugal force is balanced by gravity pulling the ball inward down the slope. If the ball is given extra speed around the wheel, motion in a straight line would appear to curve outward when viewed from the spinning wheel. In addition, the outward centrifugal force is increased by the extra speed and is no longer balanced by gravity. The two effects give an apparent force outward—the CF.

**3-SECOND BREEZE**
The Coriolis Force is perverse: whichever direction the wind is heading, the CF tries to deflect it.

**3-MINUTE SHOWER**
The Coriolis Force is important on any scale larger than the wind speed divided by the Earth's rotation about the local vertical. For a 10 ms$^{-1}$ wind this is about 124 miles. Contrary to popular belief, a bathtub is too small for the vortex above the plughole to be affected by Coriolis forces. The force was first discussed in the seventeenth century in terms of the displacement of cannonballs.

**RELATED TOPICS**
See also
THE BALANCE OF WINDS
page 40

ATMOSPHERIC WAVES
page 54

**3-SECOND BIOGRAPHY**
GASPARD-GUSTAVE CORIOLIS
1792–1843
French mathematician and mechanical engineer who discussed the forces relative to rotating waterwheels and derived an equation for CF in 1835

**30-SECOND TEXT**
Brian Hoskins

*The air appears to want to turn to the right in the northern hemisphere under the action of a fictitious force—the CF.*

# THE BALANCE OF WINDS

## the 30-second meteorology

### 3-SECOND BREEZE
Winds are like politics: to go in a straight line you have to withstand pressure from the right.

### 3-MINUTE SHOWER
The balance between winds and pressure is known as geostrophic balance, and both mid-latitude Westerlies and low-latitude Trade Winds are in this balance. Geographical changes in temperature are a prime cause of pressure variations, and there is a related balance between temperature gradients and the wind aloft, whereby a horizontal temperature gradient is associated with wind that increases with height. The temperature gradient between low and high latitudes causes Westerlies that increase with height, hence flying from North America to Europe (with the wind) is faster than vice versa.

A wind is a flow of air, commonly referring to flow on a relatively large scale. Winds blow around the Earth from west to east in mid-latitudes, and from east to west and toward the equator in the tropics, where they are called Trade Winds. Confusingly, winds are known by the direction whence they came, so Westerlies blow toward the east. Winds blow because air moves—in fact accelerates—when acted upon by a force, and pressure forces commonly arise in the atmosphere when part of the globe is heated, causing air to expand and pressure to fall. The immediate reaction of air is to flow from high pressure to low. However, the Earth is rotating, so a Coriolis Force comes into play, bending air to the right in the northern hemisphere and to the left in the southern. The upshot is that winds do not actually travel from high pressure to low; rather, they travel around regions of low pressure (cyclones) and high pressure (anticyclones) with, in the northern hemisphere, low pressure on the left and high pressure on the right.

### RELATED TOPICS
See also
PRESSURE, CYCLONES & ANTICYCLONES
page 36

CORIOLIS FORCE
page 38

### 3-SECOND BIOGRAPHIES
C. H. D. BUYS BALLOT
1817–90
Dutch meteorologist who proposed Buys Ballot's law, which states that if a person stands with his back to the wind the atmospheric pressure is low to the left, high to the right, a precursor of the notion of geostrophic balance

WILLIAM FERREL
1817–91
American meteorologist who may have preceded Buys Ballot in understanding geostrophic balance, and increased our insight into atmospheric circulation

### 30-SECOND TEXT
Geoffrey K. Vallis

*As in life, balance is key in science. The Coriolis Force is almost balanced by the pressure force; this balance lies at the very core of meteorology.*

# LOCAL WINDS

## the 30-second meteorology

## The large-scale circulation systems

which drive the weather in the extratropics are often weak or absent at low latitudes. Instead, winds are caused by topographic contrasts. Land surfaces warm and cool quickly with the day–night cycle, whereas sea temperatures remain more constant. Air heated over land becomes less dense, causing pressure to fall. This sucks denser sea air well inland on a sea breeze, causing a wind shift, temperature drop and humidity rise on its passage. The leading edge is made visible by cloud forming where air is undercut and lifted, sometimes causing showers. In the tropics it can lead to a highly predictable daily sequence. In mid-latitudes it occurs less regularly, but can be well marked in slack summertime settings, such as anticyclones. A similar mechanism causes anabatic winds, blowing up slopes warmed by the Sun, and katabatic winds blowing down slopes cooled at night. The Föhn is a class of local wind in which air forced over a mountain range is warmed and dried through condensation and rain-out of moisture. Named after an Alpine wind in southern Germany, it includes the Chinook of the North American Rockies, which in winter can cause temperature rises of 54°F in just a few hours.

**3-SECOND BREEZE**
Winds are often influenced by local topography, and can be largely driven by it, especially where large-scale pressure systems are weak.

**3-MINUTE SHOWER**
The Mediterranean has many local winds due to the complex terrain surrounding it. One such is France's Mistral, which is funneled and accelerated down the Rhône Valley and out into the Golfe du Lyon, typically when a low-pressure system forms near Genoa, Italy, a lee effect of the Alps. It can howl relentlessly for days on end and is said to affect people's state of mind, causing depression and headaches.

**RELATED TOPICS**
See also
PRESSURE, CYCLONES
& ANTICYCLONES
page 36

THE BALANCE OF WINDS
page 40

MONSOONS
page 66

**30-SECOND TEXT**
Edward Carroll

*Sea breezes and upslope, anabatic winds result from temperature, and hence pressure, differences brought about by rapid solar heating of land surfaces. Such winds focus clouds inland from the coast and over mountain tops.*

# THE GLOBAL ATMOSPHERE

**Bergen School of Meteorology**  The early realization that the large-scale behavior of the atmosphere is due to physical processes that govern the behavior of fluids—mainly hydrodynamics, thermodynamics, and mechanics—and that it can be described in mathematical terms defines an approach to meteorology that became known as the Bergen School of Meteorology. It was pioneered by a group of scientists that built up around and was influenced by Norwegian physicist Vilhelm Bjerknes, who was based at the University of Bergen in Norway. This scientific approach started to take hold in the early twentieth century immediately after the end of World War I and led to many significant developments in the science of meteorology.

**continental climate**  A climate with a large variation in temperature between the summer and winter months. These conditions occur in the interior of continents because such areas are away from the moderating influence of the sea on air temperature. This means that continental interiors tend to be hotter than coastal areas during summer and cooler than the coast in winter.

**equinox/equinoctial seasons**  The equinoctial seasons contain the equinoxes—when day and night are equally long—which occurs twice yearly, on or around March 21 and September 21. They are the seasons between winter and summer; so spring and fall are equinoctial seasons. In meteorological terms equinoctial seasons are usually considered to be the three-month periods of March, April, and May and September, October, and November. The equinoctial seasons are transition seasons between the more extreme summer and winter.

**maritime climate**  Coastal regions that benefit from a prevailing wind that comes from the sea have a maritime climate where the ocean acts to cool the air in the summer and warm it in the winter. Annual temperature variations in a maritime climate are moderate and much less than those found in continental interiors. Many western coasts of continents in mid-latitudes experience a maritime climate due to prevailing westerly winds.

**polar night/day**  At both the poles, the period when the darkness of the night lasts for 24 hours is known as the polar night and when the light of the day lasts for 24 hours it is known as the polar day. Polar days and polar nights are due to the tilt of the Earth's axis of rotation to the plane of its orbit around the Sun. It means that for a period around the summer solstice the Sun does not set and the surface of the Earth continues to benefit from warming sunlight. Similarly, for a period around the winter solstice the Sun does not rise in polar regions and this means that the surface is not warmed by sunlight and so cannot warm the air above it. The result is that temperatures drop significantly and it becomes extremely cold. The coldest-ever temperature recorded on Earth was –128°F in Antarctica during the polar night of July 1983.

**stratosphere** The layer of the Earth's atmosphere between an altitude of around 7 and 30 miles above sea level. The stratosphere begins closer to the surface at the poles (around 4 miles) and higher above the surface at the equator (around 11 miles). It features extremely cold, thin, dry air and is home to the ozone layer that protects us from much of the Sun's damaging ultraviolet light. Unlike the lower atmosphere, air temperature in the stratosphere increases with altitude due to the warming effect of this ozone, which is heated by the energy from the ultraviolet light it absorbs.

**tropopause** The boundary between the troposphere, the lowest layer of the Earth's atmosphere, and the stratosphere which begins about 7 miles above sea level. The tropopause is the border of a temperature inversion, marking the point at which temperatures stop falling with altitude in the troposphere and start rising with altitude in the stratosphere. It also acts as a boundary

between layers of the atmosphere with different chemical compositions dividing the troposphere, which contains a lot of water vapor and very little ozone, from the stratosphere, which is very dry and includes the ozone layer.

**troposphere** The lowest layer of the atmosphere where most of the weather that we experience takes place. The troposphere ranges from sea level to the edge of the stratosphere around 7 miles in altitude—higher at the equator and lower at the poles. The troposphere contains around 75 percent of the total mass of the atmosphere and nearly all its water vapor. Both the temperature of the air and the amount of moisture in the air decrease with altitude in the troposphere.

**Westerlies and Easterlies** The prevailing winds that blow from the west toward the east in the mid-latitudes of both hemispheres between 30 and 60 degrees are known as the Westerlies. Another set of winds, the Easterlies, blow from an easterly direction in a band between 30 degrees and the equatorial region of each hemisphere and are known as the Trade Winds; they blow from the northeast toward the equator in the northern hemisphere and from the southeast toward the equator in the southern hemisphere. In the polar regions above 60 degrees latitude in both hemispheres there are other easterly winds although these tend to be less regular.

# AIR MASSES & WEATHER FRONTS

## the 30-second meteorology

**3-SECOND BREEZE**
Weather fronts are sharp
transitions between air
masses—air of different
characteristics. The sharp
temperature and humidity
gradients produce clouds,
precipitation, and,
sometimes, storms.

**3-MINUTE SHOWER**
The term "weather front"
was coined shortly after
World War I, because
plotted on weather maps
these features resembled
the frontlines of armies
on military maps of the
time. The concept of
fronts was introduced
by the Bergen School
of Meteorology, which
established the principles
of how mid-latitude
cyclones form at the
interface between
cold polar and warm
subtropical air masses.

On November 11, 1911, in Springfield,
Missouri, the temperature plummeted from
80°F in mid-afternoon to 21°F by 7:00 p.m.
as cold air surged in from the northwest. This
drastic drop was accompanied by thunderstorms,
hail, and wind gusts in excess of 68 mph that
caused damage to buildings. Similar changes
were repeated across much of central U.S.,
spawning large numbers of deadly tornadoes.
Although an extreme case, what Springfield
experienced dramatically illustrates that
atmospheric temperature and humidity in
mid-latitudes tend to transition abruptly rather
than varying gradually over long distances. The
interfaces between air types are called weather
fronts, and their movement accounts for much
of the change in day-to-day weather. Between
fronts, air is more uniform and is characteristic
of its origin, leading to the idea of distinct air
masses. For example, air that has spent time
over the subpolar ocean will have different
properties to air that has been over subtropical
waters, and air masses of continental origin
are different again. Air masses meet at fronts,
where the sharp difference in temperature and
humidity can produce clouds and precipitation.
Most vigorous fronts occur as part of low-
pressure cyclones, which act to sharpen frontal
contrasts by making the opposing air masses
spiral together.

**RELATED TOPICS**
See also
PRESSURE, CYCLONES
& ANTICYCLONES
page 36

STORM TRACKS
page 52

TORNADOES
page 150

**3-SECOND BIOGRAPHY**
JACOB BJERKNES
1897–1975
Norwegian-American
meteorologist who, with
colleagues at the Bergen
School, developed the
Norwegian cyclone model

**30-SECOND TEXT**
Jeff Knight

*The atmosphere is a
restless multitude of
shifting air masses with
differences in humidity
and temperature.
Weather maps allow
air masses to be
visualized by marking
the weather fronts
where air masses meet.*

# JET STREAMS

## the 30-second meteorology

In middle latitudes the winds are generally from west to east and become stronger with height, peaking near 6 miles on the tropopause, the boundary between the troposphere and stratosphere. This region of maximum westerly winds near the tropopause is called the jet stream. Speeds are usually 40 ms$^{-1}$ (90 mph) but can be two to three times as fast. The jet stream forms a broken, wavy ribbon around the Earth, typically over 1 mile deep and 186 miles wide, but extends thousands of miles. It was discovered in the 1920s by a Japanese meteorologist, Ooishi, who observed the behavior of special balloons he released. It became well known during World War II because of its impact on aircraft, an aspect that is still important. The westerly winds increasing with height are related to the Earth's rotation and the contrast between the cold air at high latitudes and the warm air at low latitudes. The jet stream is located where the temperature contrast is strong, and so its speed is greatest in winter when this contrast is strongest. Weather systems grow on the strong temperature contrast in the jet stream region, drawing their energy from it and being steered by it.

**3-SECOND BREEZE**
The jet stream is a narrow region of high-level westerly winds that meanders around the Earth and drives weather.

**3-MINUTE SHOWER**
The Earth has one or two westerly jet streams: separate polar and subtropical jets occur during winter in the southern hemisphere and also at some longitudes in the northern hemisphere. At other longitudes they combine together to form one jet. In the summer there are easterly jets south of the Indian monsoon and over West Africa. Saturn and Jupiter have many jet streams at different latitudes because of their size and rapid rotation.

**RELATED TOPICS**
See also
LAYERS OF THE ATMOSPHERE
page 18

THE BALANCE OF WINDS
page 40

STORM TRACKS
page 52

**3-SECOND BIOGRAPHY**
WASABRO OOISHI
1874–1950
Japanese meteorologist and Director of Japan's first upper-air observatory. He wrote his 1926 report in Esperanto

**30-SECOND TEXT**
Brian Hoskins

*Such is the strength of the jet stream that it can dramatically affect flight times, depending on whether aircraft are going with or against the flow.*

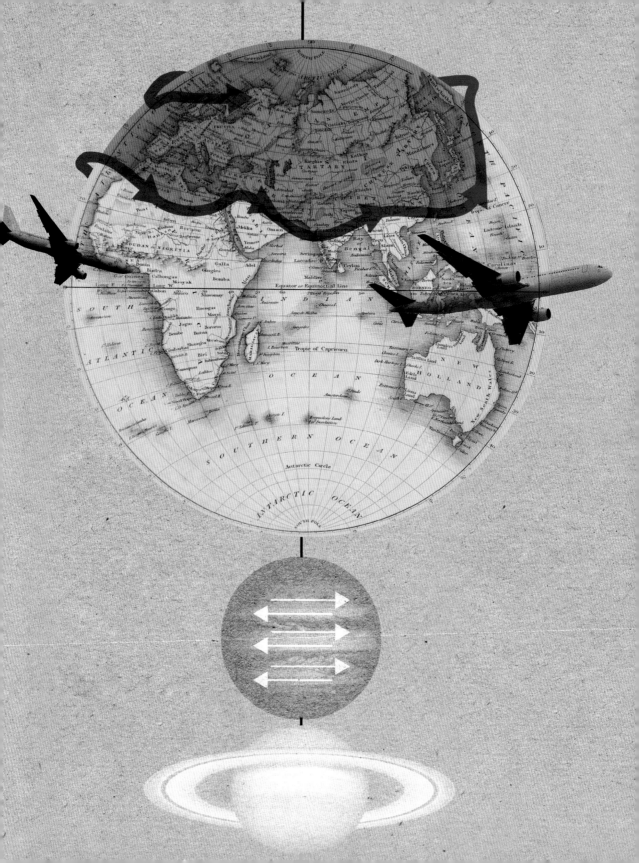

# STORM TRACKS

## the 30-second meteorology

**Storms in the middle latitudes**
typically move eastward across the major oceans
along definite paths called storm tracks. Early
sailors soon appreciated the existence of
"preferred" regions for stormy weather, and
by the middle of the eighteenth century there
were detailed maps of the North Atlantic storm
track. This begins near the east coast of North
America and usually tilts slightly northeast
across the Atlantic toward northwest Europe.
The North Pacific storm track is oriented more
west–east, from Japan to near the west coast
of North America. In the southern hemisphere
the main winter storm track spirals eastward
and poleward across the South Atlantic and
Indian Oceans, finishing close to the coast of
Antarctica. In summer it encircles Antarctica.
Storm tracks are closely related to the westerly
jet streams that cross the oceans in middle
latitudes. This is because the jet streams are
regions of strong north–south temperature
contrast, and this contrast provides much of
the energy for the growth of the storms. In
turn, the storms drive the near-surface westerly
winds that accompany the jet streams. The
storm tracks are strongest in winter when they
are also generally farthest from the poles.

**3-SECOND BREEZE**
Low-pressure weather
systems, or cyclones,
develop and move
eastward across the North
Atlantic and Pacific Oceans
on preferred paths called
storm tracks.

**3-MINUTE SHOWER**
Storm tracks can vary
in latitude and length.
Sometimes the North
Atlantic storm track
finishes near Norway,
and at other times closer
to southern Europe.
Sometimes it continues
into western Europe, and
the low-pressure systems
are young and vigorous
rather than mature and
inactive. If, however, a
blocking high pressure
dominates northern
Europe the storm track
and its associated weather
systems are unable to
get close.

**RELATED TOPICS**
See also
JET STREAMS
page 50

BLOCKING, HEATWAVES
& COLD SNAPS
page 58

**30-SECOND TEXT**
Brian Hoskins

*Storm tracks map the
typical pathways of
cyclonic storms from
their birth to their
death. Early navigators
caught in the path of
these storms were at
the mercy of the wind
and waves but modern
weather forecasting
skill means that
ships now typically
receive a week's
warning of where and
how strong such storms
are likely to be.*

# ATMOSPHERIC WAVES

## the 30-second meteorology

Waves affect everyone and everything, from waving goodbye, to the Mexican wave at a stadium, from waves on a pond's surface to devastating tsunamis. Energy comes from the Sun in lightwaves, microwaves heat our coffee, and soundwaves bring us the joys of Mozart or the Rolling Stones. Almost anything that is periodic or regular can be thought of as a wave—lightwaves are just a form of "electromagnetic" waves that oscillate trillions of times a second. The atmosphere contains waves, and not just thundering sound waves. Rossby waves circumnavigate the globe with a wavelength about the width of the Atlantic, excited by air flowing over the oceans and mountain ranges, modified by the rotation of the Earth, and they organize the weather on timescales of days and weeks. Ever wondered why the weather can get stuck in a particular pattern—rain for two weeks solid, or sunshine for days on end? Most likely it is due to Rossby waves forming a particularly persistent pattern. If we could predict these patterns better, just as surfers do for waves coming onshore, we could predict the weather farther ahead in time.

**RELATED TOPICS**
See also
CARL-GUSTAF ROSSBY
page 56

BLOCKING, HEATWAVES
& COLD SNAPS
page 58

CHAOS
page 100

**3-SECOND BIOGRAPHY**
LEONARDO DA VINCI
1452–1519
Italian artist and inventor, probably the first person to realize that sound traveled in waves, who also made some beautiful sketches of waves, among other things

**30-SECOND TEXT**
Geoffrey K. Vallis

**3-SECOND BREEZE**
Atmospheric waves try to bring order to the anarchist that is weather, by adding structure to the chaos of the circulation.

**3-MINUTE SHOWER**
Large-scale atmospheric motion is chaotic and unpredictable, but not entirely so—it is organized by planet-sized waves that periodically traverse the globe. Thus, storm tracks arise because the weather is channeled into certain regions by these waves. The weather we experience is a competition between chaotic, unpredictable storms and these more regular waves. When the waves win, we experience predictable, regular weather patterns; when chaos wins forecasts go awry. Better understanding of the atmosphere will depend on our ability to extract all the information possible from these waves.

*Waves refer to many different things but in physics, waves are oscillations—ranging from the micro to the macro scale—that transfer energy through space or mass.*

**December 28, 1898**
Born in Stockholm, Sweden

**1918**
Graduates, at the age of 19, in mathematics, mechanics, and astronomy from the University of Stockholm

**1919**
Joins the Norwegian Geophysical Institute in Bergen, Norway, to pursue an interest in meteorology

**1921**
Returns to University of Stockholm to study mathematical physics

**1923**
Publishes his first scientific paper, "On the Origin of Traveling Discontinuities in the Atmosphere"

**1926**
Relocates to America and joins the U.S. Weather Bureau in Washington DC

**1928**
Joins the newly created aeronautical engineering department at the Massachusetts Institute of Technology (MIT)

**1939**
Appointed assistant chief of U.S. Weather Bureau and becomes a U.S. citizen

**1939–40**
Authors key scientific papers that include the fundamental equation for what we now call Rossby waves

**1947**
Becomes founding director of the Institute of Meteorology in Stockholm

**1948**
Starts spending more time in Sweden where he helps to establish a national weather service

**1955**
Publishes a scientific paper reinvigorating the field of atmospheric chemistry

**1956**
*Time* magazine celebrates his contributions to meteorology in its December issue

**August 19, 1957**
Dies in Stockholm, Sweden

# CARL-GUSTAF ROSSBY

## A whole new way of thinking

about the behavior of our atmosphere was spearheaded by pioneering Swedish-American meteorologist Carl-Gustaf Rossby. By combining ideas from aeronautical engineering with the emerging field of mathematical meteorology, Rossby developed the concept of large-scale waves in the atmosphere that span a significant fraction of the Earth's circumference and gradually ripple around the planet. These waves have a major effect on the weather, particularly in mid-latitudes, and because their behavior is governed by mathematical equations linking changes in wind speed and pressure, Rossby's work contributed hugely to the development of modern computerized weather forecasting.

Rossby studied mathematics and physics in his native Stockholm before taking a job at the Geophysical Institute in Bergen at a time when atmospheric science was developing apace. He went from Bergen to Leipzig and spent much of 1921 at Lindenberg's Meteorological Observatory. To advance his understanding of upper-air data, Rossby returned to study mathematical physics in Stockholm, financing his studies by working for the Swedish Meteorological-Hydrological Service, which included serving on research expedition vessels in the North Atlantic.

In 1926 Rossby was granted a fellowship by the American-Scandinavian Foundation which took him to America, to work with the U.S. Weather Bureau, then to a post at MIT, where he established the first dedicated university meteorology course in the U.S. and set up the first-ever civil aviation weather forecasting service. His role in the aeronautical engineering department allowed Rossby to see the practical value of key physical concepts in fluid dynamics and thermodynamics.

This engineering influence resulted in two groundbreaking contributions to meteorology. The first, published in 1939, included the discovery of the equation that governs the speed of planetary-scale Rossby waves traversing through the atmosphere and shows how this is related to wind speed, latitude, and wavelength. The second, a year later, was on a quantity that is conserved in the atmosphere as air flows around, which he called barotropic vorticity. It describes how a mass of air spins as it flows around the Earth. The mathematician John von Neumann later used Rossby's equation in early computer programs for weather forecasting.

During World War II Rossby helped to organize courses for the systematic training of meteorologists in the military. After the war he spent time both in the U.S. and his Swedish homeland, collaborating widely to develop our understanding of the atmosphere including the jet stream and atmospheric Rossby waves.

Rossby spent his final years investigating the chemistry of the atmosphere but his name will forever be connected with meteorology and the global-scale waves that organize our weather.

*Leon Clifford*

# BLOCKING, HEATWAVES & COLD SNAPS

## the 30-second meteorology

A blocking high is a large high-pressure system that sits over a region such as northern Europe, often for an extended period. Its name refers to the fact that it appears to block the prevailing westerly winds and storms from the Atlantic. Typically, to the south of the high there is a low pressure and there are easterly winds between the high and the low. The high–low pressure pattern is even stronger at greater altitude, and the westerly jet stream winds split into branches going around the block far to the north and to the south. When blocking occurs and westerly winds are replaced by easterlies, the climate of western Europe is strongly influenced by the rest of the Eurasian continent, rather than by the Atlantic. In winter this produces cold dry weather and in summer hot dry weather. Because blocking often lasts for a week or more, it gives cold snaps in the winter and heatwaves in the summer. Europe is situated at the downstream end of the North Atlantic jet stream and storm track where blocking often occurs. Winter blocking also occurs near western North America at the end of the North Pacific jet stream, and west of New Zealand at the end of the Australian jet.

**3-SECOND BREEZE**
Under a blocking high, parts of the world that have a maritime climate become distinctly continental—cold in winter and hot in summer.

**3-MINUTE SHOWER**
Blocking keeps the weather systems away, but these weather systems are important for its existence. Blocking is usually initiated by a deep cyclone that slows down and moves subtropical air far poleward. This subtropical air is spinning less rapidly than the air usually in the region, and so it forms an anticyclone, or high-pressure system. After this, the weather systems approaching the block move more air poleward, which reinforces it.

**RELATED TOPICS**
See also
JET STREAMS
page 50

STORM TRACKS
page 52

**30-SECOND TEXT**
Brian Hoskins

*When a blocking high sits over a region the same kind of weather will persist for a long period—perhaps several weeks—which can be either extremely hot or distinctly chilly.*

# HADLEY CELLS
# & DESERTS

## the 30-second meteorology

During the northern hemisphere summer, there is heavy rainfall in the northern tropics, much of it associated with monsoons. This heavy rainfall is associated with large regions of thunderstorms and generally rising air. At this time, in the southern tropics and subtropics the air is descending and it is generally very dry. The rising and descending air must go somewhere, and to complete the circulation, on average there is motion from north to south high in the atmosphere and from south to north at low levels. The low-level flow is turned by the Coriolis Force and becomes the strong winter Southeast Trade Winds in the southern hemisphere. The situation is reversed during the southern hemisphere summer, which experiences rainfall and ascent, high-level motion toward the northern hemisphere, followed by descent and low-level motion back toward the southern hemisphere in the Northeast Trade Winds. In the equinoctial seasons the ascent is nearer the equator and air descends in the subtropics of both hemispheres. These average circulations in height and latitude are called Hadley Cells. At about 20–35 degrees in both hemispheres, descent and lack of rainfall dominate at most times of the year, which is why most of the world's deserts are located in these latitudes.

**3-SECOND BREEZE**
What goes up must come down—the up is in the moist tropics and the down is in the subtropical desert regions.

**3-MINUTE SHOWER**
The Hadley Cell is a picture of the average flow. However, there is much variation around the Earth. The Indian summer monsoon ascent is partially compensated by descent in the Mediterranean, resulting in its dry, hot summers. The rainfall regions in the tropical Atlantic and East Pacific stay north of the equator, even in the southern hemisphere summer, and the motions across the equator are in the opposite directions to the Hadley Cells!

**RELATED TOPICS**
See also
CORIOLIS FORCE
page 38

TRADE WINDS
page 62

MONSOONS
page 66

**3-SECOND BIOGRAPHY**
GEORGE HADLEY
1685–1768
English scientist who proposed the model for the Earth's atmospheric circulation in each hemisphere, which explained the Trade Winds

**30-SECOND TEXT**
Brian Hoskins

*The Hadley Cells are huge circulations of air in the tropics. They move moisture from the subtropics, creating most of the arid regions of the Earth, to the tropical regions of heavy rainfall.*

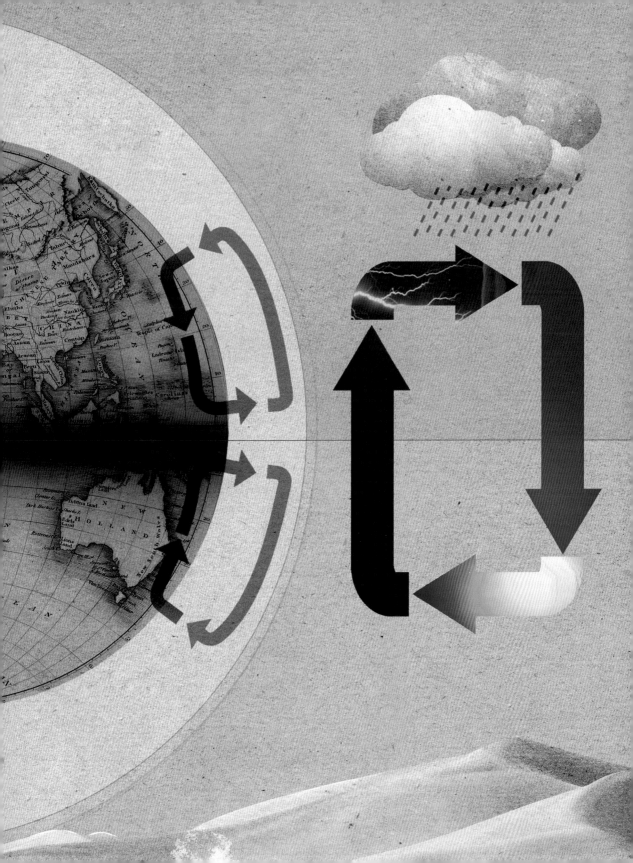

# TRADE WINDS

## the 30-second meteorology

There are three bands of wind currents in both the northern hemisphere and the southern hemisphere: in the middle latitudes, the near-surface winds blow from the west, while in the polar regions and in the tropics, the near-surface winds blow from the east. In the tropics, these steady breezes are called the Trade Winds. The trades exist out to about 30 degrees latitude in each hemisphere, and flow from the east and toward the equator, from the northeast in the northern hemisphere and from the southeast in the southern hemisphere. Even Christopher Columbus knew about the Trade Winds, because he used them to speed his journey to the New World. The Trade Winds take their strength from the rotation of the Earth. Hot air rises near the equator, and winds near the surface converge to feed the rising motion. As the winds flow toward the equator, they are bent to the right in the northern hemisphere (and to the left in the southern hemisphere) by the Coriolis force. This process gives the Trade Winds their easterly flow. The Trade Winds are strongest in winter in each hemisphere, and are characterized by a relative steadiness, compared to the disruption from weather systems that punctuate the westerly flow of the middle latitudes.

**3-SECOND BREEZE**
The steady breezes of the Trade Winds provided a reliable route for mariners to cross from Europe to the Americas.

**3-MINUTE SHOWER**
In the middle of the Trade Winds, near the equator, exist the "doldrums," a narrow band of low pressure where the winds are calm. Sailors dreaded their passage through the doldrums, because ships can drift aimlessly there for weeks, often with a dwindling supply of water and food. On the outside edge of the tropics, around 30 degrees latitude, are the "horse latitudes," another band of relative calmness.

**RELATED TOPICS**
See also
CORIOLIS FORCE
page 38

THE BALANCE OF WINDS
page 40

HADLEY CELLS & DESERTS
page 60

**3-SECOND BIOGRAPHY**
MATTHEW FONTAINE MAURY
1806–73
American oceanographer who made detailed maps of the Trade Winds and other air and ocean currents

**30-SECOND TEXT**
Dargan M. W. Frierson

*In the age of sail power the Trade Winds were a seafarer's ally but the stillness of the doldrums was the maritime equivalent of being up a creek without a paddle.*

# RAINY SEASONS

## the 30-second meteorology

When the majority of the annual rainfall for a region regularly occurs over a well-defined period of up to a few months, this period is known as the rainy season. Tropical regions, for example West Africa and Southeast Asia, tend to experience rainy seasons—usually in the summer months as monsoons—but in some regions they occur twice yearly. Tropical rainy seasons are linked to a band of clouds and thunderstorms that circle the Earth near the equator. This band, called the inter-tropical convergence zone, or ITCZ, moves, following the maximum of solar heating on the Earth's surface where the Sun's path across the sky is at its highest. This means that the ITCZ tracks around 500 miles into the northern tropics during the northern hemisphere summer and drifts into the southern tropics during the southern hemisphere summer. The heat from the Sun warms the oceans and these warm waters heat the atmosphere above. This causes the evaporation of moisture from the sea surface and strong rising motion to form clouds and thunderstorms that make up the ITCZ. The rainy season occurs when the band of ITCZ clouds and storms moves over a land mass and therefore occurs twice annually in many regions as the ITCZ shifts northward and southward.

**RELATED TOPICS**

See also
CLOUDS
page 22

RAIN
page 24

MONSOONS
page 66

**3-SECOND BIOGRAPHY**
EDMOND HALLEY
1656–1742
English astronomer who, in 1686, suggested that solar heating of the oceans was the main driver of tropical weather

**30-SECOND TEXT**
Leon Clifford

**3-SECOND BREEZE**
In some tropical regions for months it almost never rains, then it pours—the rainy seasons.

**3-MINUTE SHOWER**
The location of the migrating band of clouds and convective storms known as the ITCZ generally lags behind the relative position of the overhead Sun by one or two months. Apart from the drastic effect of the ITCZ on the frequency and intensity of rainfall over many equatorial land masses, the cumulus and cumulonimbus clouds associated with it can billow to heights of 9 miles, presenting a formidable barrier to aircraft at high altitude.

*Temperatures build up far more rapidly during the day over large land masses than oceans. As a consequence, heavy downpours in the afternoon are a characteristic feature of tropical weather.*

# MONSOONS

## the 30-second meteorology

Monsoons are caused by seasonal winds that maintain their direction for months at a time. Monsoon winds are driven by the Sun, which warms the Earth's surface. They always blow from a relatively cooler region to a relatively warmer region where the Sun-heated surface heats the air above, causing it to rise, thus drawing in more cooler air and so maintaining the wind pattern. The summer monsoon in the Indian subcontinent lasts from May to September and blows northeastward from the sea onto the hot summer land, bringing with it moist air from the southwestern Indian Ocean, which results in heavy rainfall. The Indian summer monsoon is particularly strong due to the intense heating of the land made possible, in part, because the Himalayas block cooling air from heading southward. India also experiences a weaker winter monsoon, between October and March, which sends dry air from inland China heading southwest across the subcontinent although the Himalayas act to block much of this wind from reaching the coast. The winter monsoon in Southeast Asia brings moist air from the South China Sea across Indonesia and Malaysia, causing significant rainfall. Similar wind systems occur in North and South America, northern Australia, and West Africa.

**3-SECOND BREEZE**

Monsoons are seasonal winds driven by solar heating of the Earth's surface; they are often linked with the onset of rainy seasons.

**3-MINUTE SHOWER**

Asia's monsoons provide an example of how geological processes deep within the Earth millions of years ago help to shape our weather today. Evidence of past climate conditions found in ground and ocean-floor sediments, together with computer model experiments, all suggest that the evolution of Asia's monsoons is inextricably linked with the formation of the Himalayan Mountains and the uplift of the Tibetan Plateau which began around 50 million years ago.

**RELATED TOPICS**

See also
THE BALANCE OF WINDS
page 40

TRADE WINDS
page 62

RAINY SEASONS
page 64

**3-SECOND BIOGRAPHIES**
HIPPALUS
**fl. 1st century BCE**
Greek explorer and navigator, credited by the Roman writer Pliny the Elder as being the first to document the Indian Ocean monsoon path

HENRY FRANCIS BLANDFORD
**1834–93**
British meteorologist who studied India's monsoons. He successfully predicted a monsoon rainfall failure in 1885 which caused a drought

**30-SECOND TEXT**
Leon Clifford

*Monsoon winds always blow from cold regions to warm, which, in the case of most of India and Southeast Asia, has shaped their climates.*

# STRATOSPHERIC POLAR VORTEX

## the 30-second meteorology

**3-SECOND BREEZE**
In winter a powerful circulation of air around the poles in the stratosphere of each hemisphere influences weather and the formation of the ozone hole.

**3-MINUTE SHOWER**
The stratosphere is the layer of the atmosphere above the Earth's weather; the air there is extremely dry, and most of the stratosphere has no clouds—certainly no rain. The polar vortex is so cold, however, that the tiny amount of water vapor present sometimes condenses into so-called polar stratospheric clouds. These tenuous clouds are also known as nacreous clouds, owing to their pearly, iridescent appearance near sunset or sunrise.

The fastest global-scale winds in the atmosphere are not associated with oceanic storms or America's Tornado Alley, but are high in the stratosphere between 6 and 30 miles altitude. Winds here regularly exceed 155 mph —similar to wind speeds in the strongest hurricanes. Continually whirling around the poles in wintertime, they form a gigantic cyclone known as the stratospheric polar vortex. The origin of the vortex relates to the fact that the stratosphere contains ozone, which absorbs heat from the Sun. In the polar winter, however, the Sun does not rise for months. This allows the polar stratosphere to become intensely cold—down to –120°F —far colder than the sunlit stratosphere. This temperature difference is the cause of strong winds around the polar vortex and means that the polar vortex only forms in winter. Rapid winds around the edge of the vortex also isolate air in its interior, a fact that is instrumental in the formation of the Antarctic ozone hole. In the Arctic, however, the vortex is less strong than in the Antarctic, and in some winters it is distorted or suddenly breaks down, exerting an influence on surface weather over northern polar and mid-latitudes.

**RELATED TOPICS**
See also
LAYERS OF THE ATMOSPHERE
page 18

THE BALANCE OF WINDS
page 40

JET STREAMS
page 50

THE OZONE HOLE
page 108

SUDDEN STRATOSPHERIC WARMING
page 152

**3-SECOND BIOGRAPHY**
LÉON PHILIPPE TEISSERENC DE BORT
1855–1913
French meteorologist and physicist who pioneered the use of unmanned balloons and discovered the stratosphere

**30-SECOND TEXT**
Jeff Knight

*Powerful stratospheric vortices that form over the poles in winter play a critical role in the depletion of the polar ozone layer.*

# THE SUN

**atmospheric drag** The Earth's atmosphere extends into what many of us would consider to be space and this can be seen in the effect that it has on the orbits of satellites around the Earth. The incredibly thin atmosphere at altitudes of up to 185 miles where some low Earth-orbiting satellites operate is still dense enough to hinder the movement of the satellite as it goes around the Earth. This resistance is known as atmospheric drag.

**aurora** The rippling colored lights seen in the sky on clear nights in the high latitudes of both hemispheres. The northern lights (aurora borealis) and the southern lights (aurora australis) are caused by electrically charged particles ejected from the Sun being steered into the polar regions by the Earth's magnetic field. These charged particles collide with the atoms of the air at high altitudes, which results in the atoms emitting light and so creates the aurora display. At times of high solar activity the aurora can be seen at lower latitudes.

**Brocken specter** A huge ghostly shadow cast onto cloud tops or fog and mist. It can be seen from mountain summits and ridges by climbers when the Sun is low and behind the observer. It can also be seen from aircraft flying above clouds. The specter is the shadow of the observer projected through the mist. The lack of perspective and visual reference points in cloud can create an illusion of a giant ghostly figure in the far distance. It is often associated with a halo of colors around the shadow that is centered on a point directly opposite the Sun from the observer's point of view, which can give a supernatural appearance. The name comes from the Brocken peak in Germany where this phenomenon was first recorded.

**corona(e)** Coronae are an optical effect in the atmosphere caused by thin clouds partially veiling the Sun or the Moon. They appear as concentric rings of color like a faint circular rainbow with blue innermost and red outermost. They are caused by light being diffracted as it passes through the ice crystals that make up the cloud. A lunar corona can best be seen with a full Moon.

**latitude and longitude** Latitude is the angle between a point on the Earth, the center of the Earth and the equator (the horizontal plane which divides a planet into its northern and southern hemispheres). The greater the latitude, the more northerly or southerly a point is and the closer to one of the poles. It is measured in degrees with a value between –90 or 90 degrees south (the latitude of the south pole) and +90 or 90 degrees north (the latitude of the north pole) with zero degrees being the latitude of the equator. Longitude is the angle between a point on the Earth, the Earth's axis, and the Greenwich meridian (the vertical line that defines the eastern and western hemispheres), all at the same latitude. A point can be precisely located on the Earth's surface when latitude is combined with longitude (an east–west measure).

**refraction/refractive index** Refraction causes light to bend as it passes from one material to another that has a different refractive index. The refractive index of a material is the ratio of the speed of light in a vacuum to the speed of light in that material. When light passes from one material to another with a different refractive index, it changes speed and this causes the light to bend. Density affects the refractive index of a material. Air becomes less dense with altitude and so its refractive index changes with height. This means that light passing through the atmosphere will be bent as it passes from the lower-density air at the top of the atmosphere to higher-density air near the surface. This effect can distort the appearance of the Sun and the Moon near the horizon. Turbulence in the atmosphere leading to a mixing of air with different densities causes stars to twinkle through a similar effect.

**solar constant** The energy output of the Sun is approximately constant and the Earth's orbit is nearly circular so the amount of solar energy reaching the Earth should be approximately constant too. The solar constant is the average quantity of solar energy measured in joules per second or watts falling on an area of one square meter that is perpendicular to the incidence of the Sun's rays. The solar constant has been measured to be $1.36kW/m^2$ and varies by much less than one percent—even between times of maximum and minimum solar activity.

**solar energy** Solar energy powers the weather and is the ultimate source of heat in the climate system. In meteorology, solar energy is the fraction of the Sun's energy output that is received by the Earth. Solar energy is generated within the Sun through the process of nuclear fusion and comes to Earth in the form of light and other electromagnetic radiation. X-rays and ultraviolet light emitted by the Sun are absorbed by the atmosphere. The balance of the solar energy reaching the Earth is either absorbed by the surface or reflected back into space by clouds and ice. The term solar energy is also applied to the electricity or heat generated by devices that capture this energy from the Sun.

# BLUE SKY

## the 30-second meteorology

The color of the sky varies across its expanse, from day to day, with the weather and the seasons, exhibiting a glorious array of hues. The sky does not itself emit light; the color is determined by the scattering of sunlight by the particles in the atmosphere with the amount of scattering depending on the wavelength (the color) of the light and the size of the scatterer. Shorter wavelengths, at the blue end of the sunlight spectrum, are scattered more than the longer ones, at the red end of the spectrum, so that light seen coming from the sky, having been scattered to reach the eye of the observer, has a greater component of blue. The scatterers responsible include gas molecules, suspended dust, and smoke as well as cloud water drops and ice particles. Twice the wavelength means sixteen times more scattering of smaller particles so that with a clean sky, and the scattering dominated by gas molecules, the blueness is most intense while on a hazy day, with the air full of water droplets, it can appear almost white. Approaching sunrise and sunset the red and orange colors surrounding the Sun are a result of the blue light having been scattered away.

**RELATED TOPICS**
See also
SUNSHINE
page 76

**3-SECOND BREEZE**
Air particles scatter the shorter wavelength component of sunlight more strongly so that the scattered skylight we see is bluer than the light coming directly from the Sun.

**3-MINUTE SHOWER**
Scattering by particles smaller than about one-tenth of the wavelength ($\lambda$) of light varies as $1/\lambda^4$ (Rayleigh's Law) so that blue light ($\lambda \approx 400$ nm) is scattered nearly ten times more than red light ($\lambda \approx 700$ nm) by air molecules (diameter about 0.4 nm). For dust particles of about the same size as the wavelength the scattering varies as $1/\lambda$ so the factor reduces to less than two.

**3-SECOND BIOGRAPHIES**
HORACE-BÉNÉDICT DE SAUSSURE
1744–99
Swiss physicist and Alpinist who, in 1789, invented the cyanometer based on a quantitative scale by which to measure the blueness of the sky

JOHN WILLIAM STRUTT, LORD RAYLEIGH
1842–1919
English physicist who first gave a mathematical description of scattering by small particles

**30-SECOND TEXT**
Joanna D. Haigh

*The color of the sky has provided inspiration throughout history, preoccupying painters, poets, and prophets— as well as physicists.*

# SUNSHINE

## the 30-second meteorology

### Above any location on Earth

the amount of sunlight at the top of the atmosphere is determined by latitude, time of year, and time of day. The axis of the Earth tilts so that as it marches on its annual orbit round the Sun the two poles take turns to point toward the Sun, bringing summer to that hemisphere, with the other experiencing winter. During high summer the Sun never sets at locations near the pole, and the total amount of solar energy received over a day is higher than at the Equator, while at the winter pole the days pass with the Sun never rising. The amount of sunlight at the Earth's surface depends on cloud cover and on the composition of the atmosphere: thick cloud scatters solar radiation away, and dust or other particles influence the amount reaching the ground by absorbing or scattering it. Weather stations usually host sunshine recorders, which measure the hours of sunshine during the day as well as the intensity of the solar radiation. Sunlight is fundamental to life on Earth through its role in photosynthesis but it can be harmful too: excessive ultraviolet radiation can cause genetic mutations in plants and skin cancer in humans.

**3-SECOND BREEZE**
Sunshine is a vital component of local weather: its intensity at the ground varies with the seasons and depends on cloud cover and atmospheric composition.

**3-MINUTE SHOWER**
Volcanic eruptions inject particles into the air which can significantly affect sunshine. The amelioration was so effective in 1815 following the eruption of Indonesia's Mount Tamboro that it became known as the year without a summer. Human activities also affect sunlight: industrial air pollution reduces solar radiation received at the ground while the release of chlorofluorocarbons (CFCs) created the ozone hole high in the atmosphere, resulting in a greater incidence of harmful ultraviolet radiation at the surface.

**RELATED TOPICS**
See also
SEASONS
page 20

CLOUDS
page 22

BLUE SKY
page 74

THE OZONE HOLE
page 108

**3-SECOND BIOGRAPHY**
JOHN FRANCIS CAMPBELL
1821–85
Scottish inventor of the first sunshine recorder, which consisted of a glass sphere set into a wooden bowl with the Sun burning a trace on the bowl

**30-SECOND TEXT**
Joanna D. Haigh

*A sunshine recorder provides important data for climatologists; life on Earth depends on sunshine, although overexposure can prove deadly.*

# RAINBOWS

## the 30-second meteorology

When a storm cloud, still raining, has passed overhead, and the Sun emerges behind it, you may see a rainbow in the direction of the storm. It is produced by the interaction of sunlight with the raindrops and forms an arc of a circle, the center of which lies below the horizon on the line extending from the Sun to you and onward. The rays from the rainbow make an angle of around 42 degrees to this line (no rainbow will be seen if the Sun is higher than this angle above the horizon) and at sunset the rainbow forms a semicircle. As a ray from the Sun enters a raindrop it is bent by refraction, transverses the inside of the drop, is reflected back from the far side and is bent again on its reemergence from the front. The refraction depends on the wavelength (color) of the light, so the beam is split into a spectrum of colors. The angle produced is slightly greater than 42 degrees for red light and slightly less for blue, producing the familiar color bands with red on the outside of the bow and blue on the inside. Larger raindrops produce stronger colors so that bows seen against mist are usually very pale.

**3-SECOND BREEZE**
Rainbows appear on the opposite side of the sky to the Sun; the colors are produced by the refraction of sunlight inside raindrops.

**3-MINUTE SHOWER**
A larger, secondary rainbow often appears, at an angle around 51 degrees, produced by rays which have suffered two internal reflections before emerging from the drop. This bow has its colors reversed. The sky between the two bows appears darker because the primary bow scatters some light at lesser angles, and the secondary bow some at greater angles, but not the other way round.

**RELATED TOPICS**
See also
RAIN
page 24

SUNSHINE
page 76

MIRAGES, SUNDOGS
& HALOES
page 80

**3-SECOND BIOGRAPHIES**
RENÉ DESCARTES
1596–1650
French writer, philosopher, and mathematician who gave the first quantitative description of a rainbow

ISAAC NEWTON
1642–1727
English physicist who designed an experiment to show how a rainbow forms

**30-SECOND TEXT**
Joanna D. Haigh

*Isaac Newton showed that white light is made up of all the colors of the rainbow and that refraction through a glass prism (or raindrop) separates the colors.*

# MIRAGES, HALOES & SUNDOGS

## the 30-second meteorology

### These three effects are optical

phenomena. A mirage is created when sunlight is bent by the atmosphere, causing objects to appear in unexpected places. The bending is produced by variations in the refractive index of the air near the ground through strong heating or cooling. Over hot surfaces the light is bent upward, causing an image of the sky, often resembling an area of water, to appear on the ground. Over cold surfaces the light is bent downward, sometimes causing an inverted image of a surface feature to appear in the sky. A halo around the Sun (or Moon) is seen when a veil of ice cloud extends across the sky. The ice crystals commonly take the form of small hexagonal prisms which scatter light through 22 degrees, creating a halo of this angular radius (a hand's breadth at arm's length) around the Sun. With extensive cloud cover the halo can form a complete circle but often it appears most strongly on either side of the Sun. These intense patches, called sundogs, result from the crystals orientating with their axes vertical while slowly sinking in altitude. Other bright arcs and tangents may appear, including a circle running around the sky parallel to the horizon through the Sun and sundogs.

**3-SECOND BREEZE**
The interaction of sunlight with the atmosphere creates striking and beautiful phenomena—keep an eye to the sky and marvel!

**3-MINUTE SHOWER**
Intensely colored rings sometimes appear much closer than a halo around the Sun or Moon; these coronae arise from light scattered by small water drops in the air. Looking down from a hilltop a huge shadow of the observer may be seen on a layer of mist. This is a Brocken specter, possibly accompanied by a glory (a halo of colors), with similar bright colors to a corona, surrounding its head.

**RELATED TOPICS**
See also
SUNSHINE
page 76

RAINBOWS
page 78

**3-SECOND BIOGRAPHY**
MARCEL GILLES JOZEF
MINNAERT
**1893–1970**
Flemish astronomer whose formal work concerned photometric measurements of the Sun but whose interest spanned the physics of the world around us. His volume *Light and Colour in the Open Air* (written in Dutch 1937, English translation 1954) is an inspiration

**30-SECOND TEXT**
Joanna D. Haigh

*Refraction, reflection, and diffraction of light by ice crystals, water droplets, and other material produce extraordinary optical effects in the skies.*

# SUNSPOTS & CLIMATE

## the 30-second meteorology

**3-SECOND BREEZE**

Sunspots indicate enhanced solar activity and energy output, associated with small changes in global temperature but with more significant regional impacts.

**3-MINUTE SHOWER**

Whether the energy of the Sun's radiation varied in tandem with sunspot numbers was investigated in the early 1900s using accurate radiometers sited on mountains. No consistent relationship was found and the radiant flux was generally called the "solar constant." Since the launch of satellite-borne radiometers, taking measurements outside the atmosphere, we know that solar radiation is slightly higher when the Sun has more spots.

**Sunspots are small dark features** on the Sun's surface varying in size from a few miles to many times the diameter of the Earth. Individual spots can last a few weeks and appear to move across the Sun's face as it rotates every 27 days. The total number of sunspots varies cyclically, with the length of the "11-year solar cycle" varying between 9 and 13 years. At cycle minimum there are very few spots, while at solar maximum the number can exceed two hundred. Occasionally the Sun enters a "grand minimum" (the Maunder Minimum, ca. 1645–1715, was a period of extended solar inactivity) when very few spots emerge for decades. The total energy emitted by the Sun has increased by a small fraction since the Maunder Minimum, probably associated with a small increase (< 0.18°F) in the Earth's average surface temperature since then. Regional effects may be larger with evidence that mid-latitude storm tracks shift slightly poleward when the Sun is more active, and that western Europe experiences colder than average winters during low solar activity. If the Sun declined into a new grand minimum over the next century the resultant global cooling would compensate very little for global warming, at its current rate, due to increasing concentrations of human-produced greenhouse gases.

**RELATED TOPICS**

See also
SPACE WEATHER
page 86

PAST CLIMATES
& THE LITTLE ICE AGE
page 134

MILANKOVITCH CYCLES
page 138

**3-SECOND BIOGRAPHIES**
WILLIAM HERSCHEL
1738–1822
German-British astronomer and musician, feted for his discovery of Uranus and infrared radiation, but ridiculed for his studies of the relationship between sunspots and wheat yields

JACK EDDY
1931–2009
American astronomer who suggested a relationship between solar activity and global temperature

**30-SECOND TEXT**
Joanna D. Haigh

*Sunspots, observed by the ancient Chinese and Greeks, are now monitored by satellite-mounted instruments.*

**June 14, 1868**
Born in Rochdale,
Lancashire, England

**1884**
Graduates from London
University in metallurgy

**1889**
Becomes senior wrangler
in applied mathematics at
Cambridge University and
is top of the examination
list the following year

**1904**
Becomes Director General
of Observatories in India.
Subsequently works with
the Indian Meteorological
Department on the Indian
monsoon. Elected Fellow
of the Royal Society

**1909**
Issues his first statistical
forecast of the Indian
monsoon based on
Himalayan snowfall and
recent global pressure
observations

**1911**
Becomes a companion
of the order of chivalry
known as the "Star of
India" (CSI)

**1918**
Gives the presidential
address to the Indian
Science Congress

**1924**
Publishes seminal paper,
with colleague Edward
Bliss, on world weather
correlations. Introduces
the now-popular terms
"Southern Oscillation"
and the "North Atlantic
Oscillation." Knighted
and returns to England
where he is appointed
Professor of Mathematics
at Imperial College
London

**1926–1927**
President of the Royal
Meteorological Society

**1934**
Awarded the Symons
Gold Medal of the Royal
Meteorological Society
in recognition of his
contribution to the
science of meteorology

**November 4, 1958**
Dies in Coulsdon, Surrey,
England, aged 90

**2001**
The first Sir Gilbert
Walker Gold Medal is
presented to Professor
Jagadish Shukla by the
Indian Meteorological
Society for scientific
research on the
Indian monsoon

# GILBERT T. WALKER

Gilbert Walker was born in 1868 in Victorian England in the midst of the industrial revolution. Appropriately for the time, his father was an engineer and the family moved to London where Gilbert excelled at school, his mathematical talents being particularly noted. He then graduated from London University in 1884 and went on to Trinity College Cambridge where he became a fellow in 1891. Gilbert was fascinated by the physics of spinning tops, and all kinds of flight; during his time at Cambridge he was nicknamed "Boomerang Walker," such was his fascination— he even published scientific papers on the physics of boomerangs, bird flight, and sports and games including golf and billiards.

Gilbert became Director General of the Indian Meteorological Department in 1904 and quickly realized that previous long-range forecasts of the Indian monsoon were not based on robust results. His views were summed up in a later quote: "I think that the relationships of world weather are so complex that our only chance of explaining them is to accumulate the facts empirically . . ." So Gilbert set about systematically collecting all of the observational data for the monsoon and worldwide climate that he could obtain. Along with his assistants he created a statistical forecast model for the Indian monsoon based on Himalayan snowfall and worldwide pressure observations and issued his first Indian monsoon forecast in 1909.

World War I slowed progress in subsequent years. During this time Gilbert coordinated numerous Indian Meteorological Department staff to form a "human computer" in order to perform further statistical calculations on the Indian monsoon and worldwide weather—an early forerunner of the numerical computer simulations carried out by meteorologists today.

In 1924 Gilbert and a colleague published what is perhaps his most remembered contribution to meteorology. The distillation of many years of weather observations and calculations of relationships between global weather patterns led to his description of the major patterns of global climate variability. He introduced the terms "Southern Oscillation" and "North Atlantic Oscillation." He also described the deep atmospheric flow in the equatorial Pacific now known as the "Walker Circulation." The foundations laid by Gilbert Walker earned him a knighthood and led to subsequent breakthroughs in long-range forecasting using the Southern Oscillation and its relation with El Niño and along with the North Atlantic Oscillation his work still inspires intense scientific research.

*Adam A. Scaife*

# SPACE WEATHER

## the 30-second meteorology

**3-SECOND BREEZE**
Solar storms produce
bursts of charged
particles and radiation
which reach the Earth,
affecting its atmosphere
and modern technology.

**3-MINUTE SHOWER**
An emerging activity in
forecasting space weather
aims to mitigate its effects.
Solar particles take from a
few minutes to a few days
to reach Earth so that
spacecraft positioned
between the Sun and the
Earth can provide enough
advance warning, at least
of slower-moving events,
to allow sensitive systems
to be switched off or
protected. Computer
models of the upper
atmosphere are also
being developed to
predict impacts.

Throughout history, auroral displays in the skies at high latitudes have been a source of wonder but not until the twentieth century were they understood to be caused by particles emanating from the Sun. The electrically charged particles form the solar wind that constantly bathes the Earth and which, due to interaction with the Earth's magnetic field, reaches farther into the atmosphere near the poles. Variations in the solar wind are produced by solar storms, flares, and particle ejections, which are intermittent but tend to be more frequent when the Sun approaches the peak of its 11-year sunspot cycle. Such events exert various impacts on the Earth's environment, which are generally termed space weather. At the surface they cause changes in the Earth's magnetic field, apparent as variations in the direction of compass needles, and produce electrical currents resulting in problems with transmission lines and transformers. Sudden changes in the magnetic field influence electrical currents flowing in the upper atmosphere and thereby impact on the transmission of long-distance radio signals and signals from communication and GPS satellites through malfunctions in electronics. Pilots and astronauts are subject to enhanced radiation hazard while variations in solar heating affect atmospheric drag and thus spacecraft orbits.

**RELATED TOPICS**
See also
SUNSPOTS & CLIMATE
page 82

**3-SECOND BIOGRAPHIES**
RICHARD CARRINGTON
1826–75
British amateur astronomer who, in 1859, made the first observation of a solar flare and proposed a connection with the geomagnetic storm measured on Earth the following day

EUGENE PARKER
b. 1927
American astrophysicist who proposed the existence of the solar wind

**30-SECOND TEXT**
Joanna D. Haigh

*The Sun is an active star and its periodic unleashing of solar particles can produce visual effects and geomagnetic storms and cause disruption to power grids, navigation, and communication systems on Earth.*

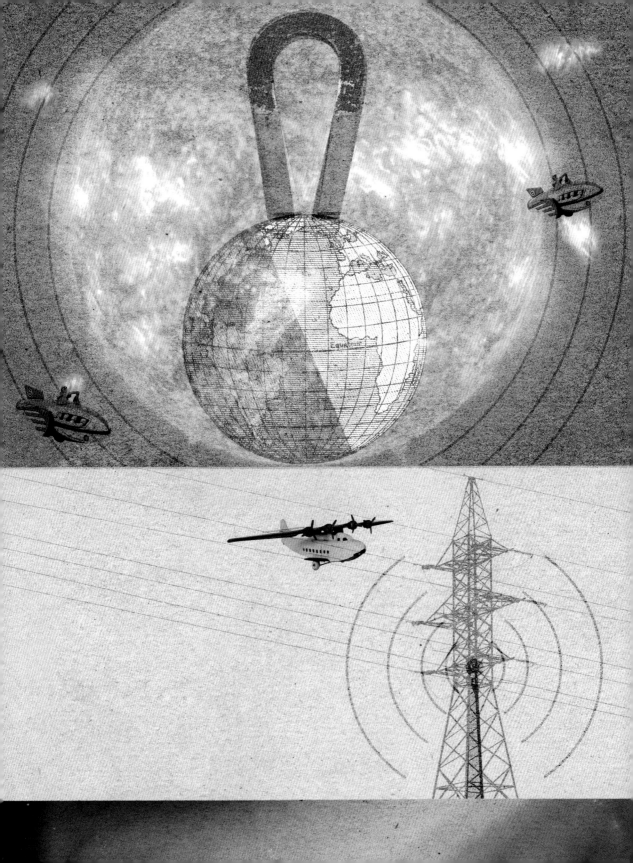

# WEATHER WATCHING & FORECASTING

**electromagnetic radiation**  A form of energy that is described in terms of vibrations in electric and magnetic fields. Most of the energy emitted by the Sun is in the form of electromagnetic radiation. It travels at the speed of light and transports solar energy from the Sun to the Earth. The different types of electromagnetic radiation are distinguished by their wavelength (or color) and by their frequency; the shorter the wavelength the higher the frequency, and vice versa. The smallest unit of electromagnetic radiation is the photon. Photons of light with shorter wavelengths, and higher frequencies, carry greater energy than photons of longer wavelength light. Different atmospheric gases transmit and absorb electromagnetic radiation of various wavelengths in different ways. Carbon dioxide is transparent to visible light but absorbs infrared and ozone absorbs ultraviolet light but transmits visible light. When an atmospheric gas absorbs electromagnetic radiation, it acquires the energy in that radiation and so warms.

**general circulation models (GCMs)**
These mathematical models are used to simulate the Earth's weather and climate. They treat the atmosphere and the oceans as fluids on a rotating sphere that behave in accordance with the laws of physics, and, in particular, the equations of mechanics, hydrodynamics, and thermodynamics. They perform many millions of calculations per second. There are both atmospheric and oceanic GCMs and these can be coupled to simulate the combined system of the atmosphere and the oceans. GCMs form the basis of modern computerized weather forecasting and also climate predictions.

**longwave radiation**  The heat radiated by the warm surface of the Earth and warm areas in the atmosphere. It has a longer wavelength than the visible and ultraviolet light from the Sun, which is known as shortwave radiation. Longwave radiation is infrared and invisible but it is a form of electromagnetic radiation just like light and radiowaves. The Earth's surface absorbs incoming sunlight which results in the surface heating up and leads to the emission of this infrared light. Some of this heat radiated by the Earth is absorbed by clouds and then reradiated both upward and downward. The downward portion helps to keep the surface warm on cloudy days. Some is absorbed by carbon dioxide and other greenhouse gases in the atmosphere and is also reradiated both upward and downward; this is the greenhouse effect that warms the planet. The fraction of the radiation emitted by the surface that eventually escapes into space is known as outgoing longwave radiation.

**microwave radiation** Microwaves are a form of electromagnetic radiation. They are high-frequency, short wavelength, radio waves with a frequency range of between 300MHz and 300GHz. Microwaves above 20GHz are absorbed by water droplets in the air, water vapor, and other atmospheric gases. Oxygen in the atmosphere emits microwave radiation; the greater the temperature of the atmosphere, so the greater the intensity of the microwave emissions from the oxygen in the air. This phenomenon allows satellites to measure the temperature of the atmosphere using microwave-detecting instruments.

**numerical weather prediction (NWP)**
The use of mathematical equations, which encode the laws of physics operating on the atmosphere, to predict how the atmosphere will change over time from its current state. In principle, these calculations could be done by humans but the huge number of calculations required to generate a meaningful weather forecast means that computers and GCMs (see facing page) are needed. NWP was conceived by meteorologist Lewis Fry Richardson after World War I. The first 24-hour NWP forecast by computer was produced in 1950 by the U.S. Army's ENIAC computer.

**stratosphere** The layer of the Earth's atmosphere between an altitude of around 7 and 30 miles above sea level. The stratosphere begins closer to the surface at the poles (around 4 miles) and higher above the surface at the equator (around 11 miles). It features extremely cold, thin, dry air and is home to the ozone layer that protects us from much of the Sun's damaging ultraviolet light. Unlike the lower atmosphere, air temperature in the stratosphere increases with altitude due to the warming effect of this ozone, which is heated by the energy from the ultraviolet light it absorbs.

**troposphere** The lowest layer of the atmosphere where most of the weather that we experience takes place. The troposphere ranges from sea level to the edge of the stratosphere around 7 miles in altitude—higher at the equator and lower at the poles. The troposphere contains around 75 percent of the total mass of the atmosphere and nearly all its water vapor. Both the temperature of the air and the amount of moisture in the air decrease with altitude in the troposphere.

# WEATHER RECORDS

## the 30-second meteorology

## 3-SECOND BREEZE
Records of the weather are of many types but common are temperature, rainfall, snowfall, and wind speed and direction.

## 3-MINUTE SHOWER
The earliest weather records, now lost, were probably those of rainfall used for taxing agricultural output in India around the fourth century BCE. The earliest extensive weather diary known was written by a cleric, William Merle, in 1337–44 in England. The longest-running weather record is that of temperature in the English Midlands, which began in 1659. Other places with very long continuous weather records include De Bilt (Holland), Stockholm (Sweden), and Philadelphia (USA).

Weather records can be both measurements and descriptions of meteorological events. Once only collected in paper form, today weather records are often merged into large digital databases. To be truly useful, they must be carefully corrected to remove errors that may derive from the original measurements or codes used to exchange data. Weather records are vital for studying climate—the statistics of weather averaged over decades or more—its variability and changes. Their other main uses are to provide starting data for weather forecasts, which form the basis of many services required by businesses and governments for operations and planning. In 1853 an International Conference created the current international arrangements for measuring weather data over the oceans, motivated by increasing losses from storms as trade expanded. Modern weather records are derived from instruments situated on the ground, on buoys and ships stationed in the oceans, and from orbiting and geostationary satellites. These instruments' data, some continuously recorded, are exchanged in near-real time in a specially coded form between nations via the Global Telecommunication System organized by the World Meteorological Organization, a U.N. agency, as well as in delayed form through international publications such as *World Weather Records* or the internet.

**RELATED TOPICS**
See also
PRESSURE, CYCLONES
& ANTICYCLONES
page 36

WEATHER FORECASTING
page 98

**3-SECOND BIOGRAPHIES**
MATTHEW FONTAINE MAURY
1806–73
American naval officer who organized an International Conference in Brussels in 1853 to coordinate wind and other atmospheric and ocean surface measurements over the oceans

GORDON MANLEY
1902–80
English climatologist who created the Central England Temperature record, the world's longest continuous weather record, starting in 1659

**30-SECOND TEXT**
Chris K. Folland

*From early surface observations to the latest satellite data, weather records are crucial for monitoring weather and for starting forecasts.*

**January 1, 1917**
Born in San Francisco,
California

**1940**
M.Sc. in mathematics
awarded by UCLA

**1946**
Ph.D. in meteorology
awarded by UCLA

**1946**
Spends a year working
with Carl-Gustaf Rossby
at the University of
Chicago

**1946**
Meets John von Neumann

**1947**
Publishes major paper
showing impact of global
temperature differences
on atmospheric waves

**1948**
Publishes major paper:
"On the Scale of
Atmospheric Motions"
featuring the
quasigeostrophic
vorticity equations

**1948**
Joins Institute of
Advanced Study,
Princeton, to work with
von Neumann's Electronic
Computer Project

**1949**
Publishes major paper:
"On a Physical Basis for
Numerical Prediction of
Large-Scale Motions in
the Atmosphere"

**1950**
Becomes part of the
team that produced the
first computer weather
forecast using the
ENIAC machine

**1956**
Becomes Professor of
Meteorology at MIT

**1957**
Appointed to the
U.S. National Academy
of Sciences' Commission
on Meteorology

**1979**
Chairs a committee
that produced a report
on the likely impact of
carbon dioxide on climate

**June 16, 1981**
Dies in Boston,
Massachusetts

# JULE CHARNEY

**It took a meteorologist who had** trained as a mathematician to make the key breakthrough that made possible computerized weather forecasts: the American scientist Jule Charney.

Born in San Francisco in 1901, Charney was a gifted mathematician at school who went on to study physics and mathematics at the University of California in Los Angeles (UCLA). He developed an interest in meteorology as the topic for his doctoral thesis following the establishment of a meteorology group at UCLA.

Charney built on the earlier work of British meteorologist Lewis Fry Richardson who had used the differential equations of hydrodynamics that govern how wind speed and air pressure change over time to attempt a numerical weather forecast. Charney was able to modify these hydrodynamical equations by extracting the key meteorological terms from them. The result was the quasigeostrophic vorticity equations—a set of simplified hydrodynamical equations for calculating large scale motions in the atmosphere that would make numerical weather forecasting much more straightforward.

During World War II, and while working on these ideas, Charney became involved in the wartime effort to train military meteorologists and in this role he made contact with many of the leading meteorologists of the day. There then followed a phenomenally productive spell: a 1947 paper based on his doctoral thesis showed how north-south temperature differences affect waves in the atmosphere; in 1948 he published his simplified hydrodynamic calculations; and in 1949 he showed how these would form the basis for computerized weather forecasts.

Subsequently, Charney started working with computer pioneer John von Neumann at Princeton on a project to apply the emerging technology of electronic computers to weather forecasting. Charney's modified hydrodynamical equations were ideally suited for computers and Charney was part of von Neumann's team that programmed the U.S. Army's ENIAC computer to produce the first-ever one-day numerical weather forecast in 1950.

Charney's equations became the basis for the first general circulation models (GCMs) of the atmosphere—the forerunners of the powerful computer climate models used by meteorologists and climate scientists today.

Charney went on to contribute hugely to organizing meteorology in the U.S. and to chair a group of scientists that produced a major report in 1979 summarizing the potential impact of increasing levels of carbon dioxide on the atmosphere. However, his greatest achievement was that he made it possible for computers to generate our weather forecasts.

*Leon Clifford*

# WEATHER SATELLITES & RADAR

## the 30-second meteorology

All matter emits energy in the form of electromagnetic radiation. As temperature increases, the amount of radiated energy increases but the wavelength of peak emission decreases. Solid matter has a continuous spectrum of emission but gases are selective in the wavelengths of radiation they absorb and emit, and it happens that the atmosphere is largely transparent at both the Sun's and the Earth's wavelengths of peak emission. Satellite-borne instruments can therefore measure visible, solar radiation scattered back by clouds, land, or sea, as well as longerwave, infrared radiation emitted by those surfaces, to build a complete picture, much as a camera does. However, by also sensing a series of wavelengths at which the atmosphere is varyingly opaque, satellites "see" different levels of the atmosphere itself, detecting radiation directly emitted by gases, and thereby indirectly measuring the atmospheric temperature. Humidity can also be estimated by selecting wavelengths preferentially emitted by water vapor, and winds at different levels are inferred by tracking clouds. Weather radars are groundbased instruments that emit pulsed, microwave radiation, which is scattered back by rain in proportion to the number and diameters of the drops. The delay and strength of the return echoes allow rainfall location and intensity to be gauged.

**3-SECOND BREEZE**
The atmosphere is under intensive scrutiny from orbiting satellites and groundbased radars, which provide a three-dimensional picture of its continuously changing state.

**3-MINUTE SHOWER**
Successful prediction of the weather for days ahead relies on accurate and detailed representation of today's global atmosphere by computer models. Measurements from satellite-borne instruments, the most important source of this information, have delivered substantial increases in forecast accuracy. Radar networks showing the movement of rainbearing systems are key to accuracy of 0–6-hour precipitation forecasts—handy, for example, in deciding whether to leave the washing out.

**RELATED TOPICS**
See also
LAYERS OF THE ATMOSPHERE
page 18

WEATHER FORECASTING
page 98

**3-SECOND BIOGRAPHY**
MAX PLANCK
1858–1947
German physicist, most famous for the quantum theory, but also the discoverer, in 1900, of the Planck function

**30-SECOND TEXT**
Edward Carroll

*Meteorological satellites either orbit the Earth from pole to pole, with a changing field of view of the rotating Earth beneath, or orbit with it, keeping a fixed vantage point high above the equator. Planck's function enables the radiation measured by satellites to be converted into the temperatures of matter below.*

# WEATHER FORECASTING

## the 30-second meteorology

Since the seminal work of Abbe (1901), Bjerknes (1904), and Richardson (1922), the challenge of forecasting has been formulated as an initial value problem of mathematical physics based on equations of geophysical flows. By approximating the equations using numerical methods, the problem of prediction has been solved. The success of the first numerical computer prediction by Charney, Fjortoft, and von Neumann in 1950 launched a formidable trend of research, development, and operational applications in Numerical Weather Prediction. These major advances influenced other branches of science, like the work of Lorenz in the 1960s on chaos. The increase in forecast accuracy during the last few decades was the result of a sophisticated interplay of advancements in numerical methods, the mathematical treatment of physics (clouds, mountains, turbulence, radiation, etc.), improved treatment of surface and space observation systems, and computing systems. Increased computer power drives weather science by allowing increased space–time resolution in the modeling of the atmosphere's dynamical and physical processes. Each order of magnitude increase in computing power improves accuracy, with huge and diverse economic impacts, ranging from emergency response for high-impact weather such as hurricanes, floods, and snowstorms, to management of hydro- and windpower production.

### RELATED TOPICS
See also
LEWIS FRY RICHARDSON
page 34

JULE CHARNEY
page 94

CLIMATE PREDICTION
page 102

EDWARD NORTON LORENZ
page 148

### 3-SECOND BIOGRAPHIES
CLEVELAND ABBE
1838–1916
American meteorologist and a founder of the National Geographic Society. He devised weather prediction based on Eulerian methods published in 1901 as "The Physical Basis of Long Range Weather Forecast"

VILHELM BJERKNES
1862–1951
Norwegian meteorologist who established the Bergen School of Meteorology. Craters on the Moon and Mars are named Bjerknes in his honor

### 30-SECOND TEXT
Gilbert Brunet

*Forecasting has changed from an art to a science in less than 50 years.*

**3-SECOND BREEZE**
The five-day weather forecast today is as accurate as the one-day forecast from 40 years ago.

**3-MINUTE SHOWER**
Forecasting is evolving toward environmental prediction that will permit us to forecast not only the atmosphere but the ocean, the full water cycle, and atmospheric composition. The weather-forecasting problem is increasingly more multiscale and complex. This brings formidable prediction challenges, including self-organized cloud systems, chaotic behavior at different space–time scales, probabilistic estimation of forecast uncertainties, forecasting rare events, and better understanding of ecosystem responses to changes in geophysical parameters.

# CHAOS

## the 30-second meteorology

**From mathematic considerations,** French polymath Henri Poincaré concluded in 1908 that a small meteorological observation error could grow into a huge error later in the forecast. During the 1960s Edward Lorenz used computers and simple numerical models to rigorously study this predictability problem and its "sensitivity to initial conditions" and established chaos theory. Through this seminal work, today's meteorologists understand that the atmosphere is chaotic, and that, while a storm will develop somewhere, it is sometimes not possible to be precise about where and when. In modern weather forecasting, uncertainties due to chaos are estimated by computing multiple forecasts, each having minute differences in the forecast computer model and initial observations. As described by chaos theory, these perturbations amplify over time in local weather features, and eventually evolve into plausible but different future states of the weather. The benefit provided by these multiple forecasts is analogous to group decisionmaking compared with the opinion of one individual. In practice, meteorologists try to reliably predict the probability of specific weather events. We can now predict with confidence high-impact weather—like hurricane landfall, ground frosts, or precipitation— with useful probabilities 3–6 days ahead.

## RELATED TOPICS
See also
WEATHER SATELLITES
& RADAR
page 96

WEATHER FORECASTING
page 98

CLIMATE PREDICTION
page 102

EDWARD NORTON LORENZ
page 148

## 3-SECOND BIOGRAPHY
JULES HENRI POINCARÉ
1854 –1912
French mathematician, physicist and philosopher of science, and the first to describe the behavior of a chaotic system. He is considered one of the last authentic polymaths of the twentieth century

## 30-SECOND TEXT
Gilbert Brunet

**3-SECOND BREEZE**
Perfect prediction of the weather is impossible in a world of deterministic chaos and meteorologists need multiple forecasts to predict the risk of different weather events.

**3-MINUTE SHOWER**
Chaos theory tells us that differences between two near-identical weather forecasts increase exponentially at first, but physical constraints such as the conservation of total energy mean they cannot diverge forever. The predictability limit is reached when the difference makes the forecast useless. In general predictability limits range from a few hours for local weather to much longer for continental-scale weather.

*An open mind and a probabilistic approach are required to deal with chaos in weather and climate forecasts.*

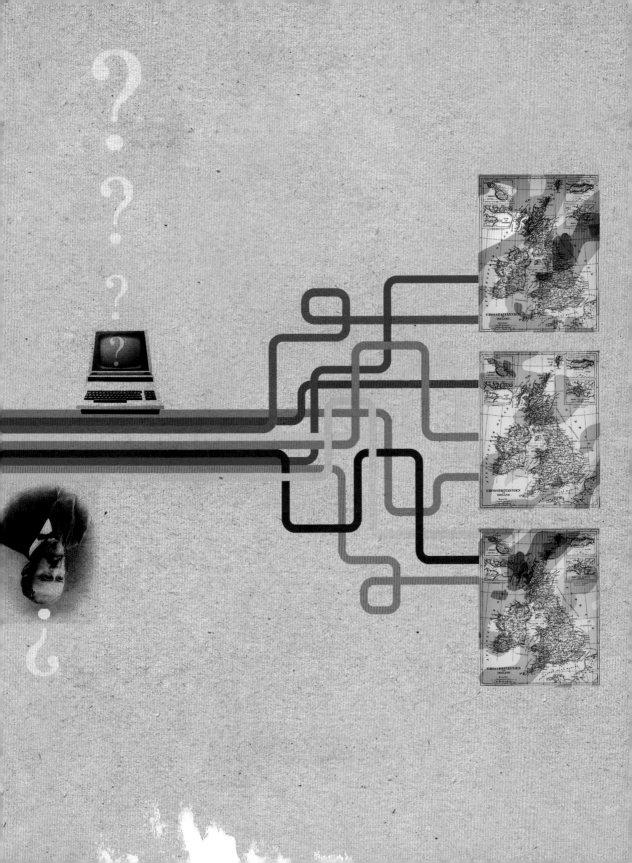

# CLIMATE PREDICTION

## the 30-second meteorology

Early attempts at predicting the climate mainly involved the fruitless search for simple cycles in meteorological data. Modern climate prediction followed the advent of computer weather forecasting in the 1960s. Weather-forecasting computer models are applied to predict climate by simulating weather across the globe, hour by hour for years or decades ahead. Even early models supported the theoretical idea that continued emissions of greenhouse gases would lead to significant global warming of climate and, as a result, this has become the central focus of climate prediction. In the following decades, models have become more sophisticated by building in mathematical representations of other climate components, such as oceans, land surface, and icecaps, as well as chemical and biological processes. These allow better, more detailed estimates of climate change, including the effect of glacier melting on sea-level rise. The effectiveness of climate models has been demonstrated many times, for example, by reproducing global climate changes in the last century or predicting the cooling following the 1991 volcanic eruption of Mount Pinatubo in the Philippines. Despite their success, challenges remain, such as improving predictions of changes in regional mid-latitude weather patterns.

**3-SECOND BIOGRAPHY**
SYUKURO MANABE
1931–
Japanese climatologist who developed one of the earliest climate model simulations of the effect of increasing greenhouse gases

**30-SECOND TEXT**
Jeff Knight

*Climate models reproduce many features of our current climate and they all predict a warmer world for the future.*

# CAN WE CHANGE THE WEATHER?

**blackbody radiation** All objects emit electromagnetic radiation. The amount and mix of wavelengths of this radiation depend on the temperature of the emitting object. Blackbody radiation is the name given to the spectrum of electromagnetic radiation that would be spontaneously emitted by a perfectly black body that is in thermodynamic equilibrium with its environment; that is, at a stable temperature. This radiation is emitted over a range of wavelengths and the proportion of shorter-wavelength radiation in the mix grows as temperature rises. At higher temperatures objects will begin to glow red, then yellow and white—hence the expression white hot. The spectrum of radiation an object emits is used to estimate its temperature. At room temperature the majority of blackbody radiation is infrared and cannot be seen. The temperature of the Earth's surface is around room temperature, which is why the Earth naturally radiates invisible infrared or longwave radiation.

**chlorofluorocarbons (CFCs)** Humanmade chemicals that have led to the depletion of the ozone layer that protects the surface of the Earth from the Sun's harmful ultraviolet rays. They are also powerful greenhouse gases that were added to the atmosphere as a result of human industrial activity. CFCs are nontoxic, chemically inert, and industrially useful. They have been widely used as refrigerator and air-conditioner coolants, aerosol spray-can propellants, and as solvents. There are many different variations of CFC molecules and they are categorized with a numbering system. They all contain atoms of carbon, chlorine, and fluorine. In the upper atmosphere, CFC molecules become exposed to ultraviolet light from the Sun, which causes them to break down, releasing chlorine, which then acts to catalyze chemical reactions that destroy ozone. A single chlorine atom released from a CFC molecule can catalyze reactions that destroy some 100,000 ozone molecules.

**greenhouse gases** The atmosphere keeps the surface of the Earth much warmer than it should be, considering the distance of the Earth from the Sun, thanks to the greenhouse effect. Visible light from the Sun, known as shortwave radiation, passes through the atmosphere and is absorbed by the land and the sea, causing them to warm. The Earth's warm surface emits this heat as infrared light, also known as longwave radiation, which is absorbed by carbon dioxide, methane, water vapor, and other so-called greenhouse gases in the atmosphere. These gases reradiate this absorbed heat both upward and downward and the downward portion warms the surface below. Humanmade global warming is caused by human activity adding additional carbon dioxide and other greenhouse gases to the atmosphere and so strengthening the natural greenhouse effect.

**Montreal Protocol** A global treaty aimed at reducing the damage caused by CFCs to the ozone layer. Signed in 1987, the Montreal Protocol to Reduce Substances that Deplete the Ozone Layer has subsequently been universally ratified. It set out to halve 1986 production levels of CFCs before 2000; it has since been amended and tightened with a view to eliminating the use of these chemicals. Research suggests that atmospheric CFC levels have fallen as a result and that the ozone hole over the Antarctic has started to recover.

**precipitation** When water vapor from the atmosphere condenses and falls to the ground under the influence of gravity this is known as precipitation. It includes rain, snow, sleet, and hail, which all fall downward, but not fog and mist because the condensed water vapor that constitutes fog and mist remains suspended in the atmosphere rather than falling to the ground. Precipitation is triggered when water vapor in the air condenses. This occurs when the air becomes saturated, usually as a result of falling temperatures, which reduce the capacity of the atmosphere to contain water vapor.

**smog** A type of air pollution that results in a thick smoky fog near the ground. The word smog is a combination of fog and smoke. It can be made worse when a temperature inversion in the atmosphere traps pollution and allows the smog to accumulate. Smog includes soot particles from smoke as well as other pollutants such as vehicle exhaust gases. It is a feature of large urban areas and was a particular problem in London, England, until the 1950s as a result of coal burning. Today, vehicle emissions cause smog in many cities. Widespread burning of agricultural and forest land in Asia also causes smog problems.

**stratosphere** The layer of the Earth's atmosphere between an altitude of around 7 and 30 miles above sea level. The stratosphere begins closer to the surface at the poles (around 4 miles) and higher above the surface at the equator (around 11 miles). It features extremely cold, thin, dry air and is home to the ozone layer that protects us from much of the Sun's damaging ultraviolet light. Unlike the lower atmosphere, air temperature in the stratosphere increases with altitude due to the warming effect of this ozone, which is heated by the energy from the ultraviolet light it absorbs.

# THE OZONE HOLE

## the 30-second meteorology

Life on Earth depends on the ozone layer, which is located in the stratosphere around 12 miles high. Ozone blocks powerful ultraviolet light, which causes DNA mutations and leads to skin cancer and cataracts. In 1974, researchers discovered that a group of common synthetic chemicals could result in widespread destruction of ozone. Chlorofluorocarbons (CFCs) were being used in great quantities in spray cans, refrigerators, and air-conditioners, and were building up in the atmosphere. A single CFC molecule, after rising into the stratosphere, can destroy over 100,000 ozone molecules. In 1985, from their station in frigid Antarctica, scientists made one of the most surprising discoveries of the twentieth century: a giant hole was appearing in the ozone layer each spring. Chemical reactions on the surface of nacreous clouds are key in causing the rapid decline. The ozone hole is a vivid symbol of the dramatic effect humans can have on the environment. Soon after the discovery of the ozone hole over Antarctica, governments agreed to ban CFCs with the signing of the Montreal Protocol in 1987. Signed by every country in the world, the Protocol has been one of the most successful environmental agreements in history.

### RELATED TOPICS

See also
LAYERS OF THE ATMOSPHERE
page 18

STORM TRACKS
page 52

POLAR VORTEX
page 68

SUNSHINE
page 76

### 3-SECOND BREEZE
Chemicals in hairsprays and refrigerators brought the life-protecting ozone layer to its knees, opening a giant hole over Antarctica that lets in dangerous ultraviolet light.

### 3-MINUTE SHOWER
Because the chemicals that destroy ozone are so persistent, the ozone hole has only recently started to recover. Damage will persist until around 2060. The ozone hole has recently been linked to a large shift in the strongest storms on the planet. The southern hemisphere storm track, located south of Australia and Africa, has shifted south and strengthened in response to the ozone hole, bringing its high winds and rainfall with it.

### 3-SECOND BIOGRAPHIES
JOSEPH FARMAN
1930–2013
British scientist who discovered the ozone hole over Antarctica

MARIO MOLINA
1943–
Mexican atmospheric chemist and Nobel Prize winner for linking CFCs and ozone depletion

### 30-SECOND TEXT
Dargan M. W. Frierson

*Synthetic chemicals cause damage to the ozone layer. Discovery of a massive hole over Antarctica in 1985 prompted global action.*

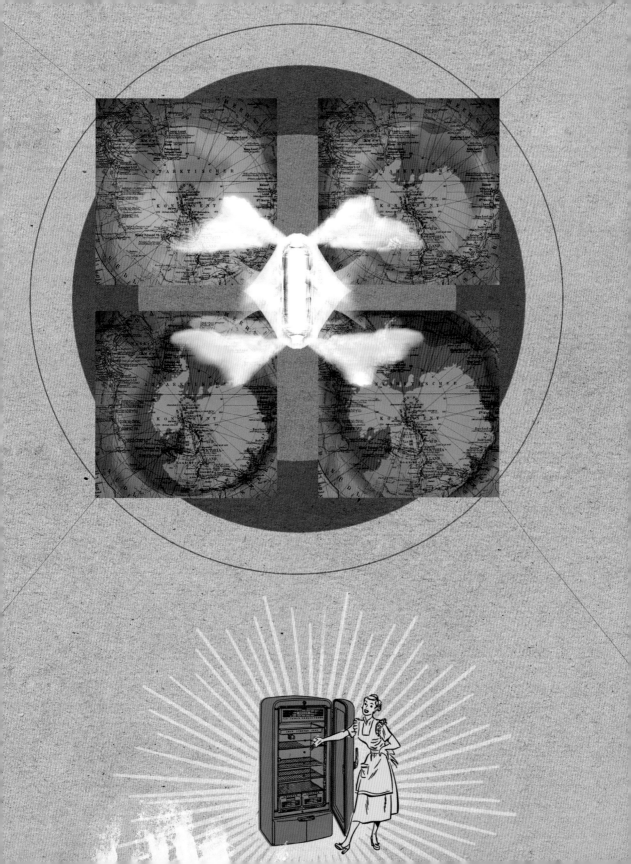

# GLOBAL WARMING & THE GREENHOUSE EFFECT

## the 30-second meteorology

**3-SECOND BREEZE**
The climate is rapidly changing, we're to blame, and it's going to get much worse.

**3-MINUTE SHOWER**
Serious harm to the climate can be prevented by making aggressive cuts to greenhouse gas emissions, for instance by switching to wind and solar power, electric vehicles, and even favoring lower-beef diets, since cattle produce methane. Were emissions to drop rapidly, we could limit the additional warming in the coming century. However, emissions continue and hence atmospheric greenhouse gases continue to rise.

Without greenhouse gases in our atmosphere, the Earth would be a frigid ball of ice. Water vapor, carbon dioxide, methane, and other greenhouse gases make it harder for heat to escape to space, making our planet warm and habitable. But since the Industrial Revolution, human activities have caused a rapid rise in such gases; in response, our planet is overheating. Burning fossil fuels and deforestation have led to a 40 percent increase in atmospheric carbon dioxide—the primary cause of global warming of nearly 2°F recorded over the last century. Climate change is most rapid in the Arctic, where sea ice over the North Pole has thinned and retreated dramatically in the last few decades, allowing more warmth to escape from the ocean. Worldwide, sea levels are rising and heatwaves are more frequent. A warmer atmosphere contains more water vapor, which supercharges weather systems, making extreme rainfall events more common. Meanwhile, persistent drought is likely to continue to plague already dry regions like the Mediterranean and south Australia. In coming decades, if emissions keep rising, warming is set to accelerate, with several degrees of temperature increase expected by 2100 and largest changes over land and the Arctic. Such extreme climate changes would present major environmental, economic, and social challenges.

**RELATED TOPICS**
See also
CLIMATE PREDICTION
page 102

SVANTE ARRHENIUS
page 112

PAST CLIMATES
& THE LITTLE ICE AGE
page 134

**3-SECOND BIOGRAPHIES**
JOSEPH FOURIER
1768–1830
French mathematician and discoverer of the greenhouse effect

CHARLES DAVID KEELING
1928–2005
American atmospheric chemist and the first to measure the accumulation of carbon dioxide in the atmosphere

**30-SECOND TEXT**
Dargan M. W. Frierson

*Unless checked, global warming will have significant and harmful effects on agriculture, water resources, and ecosystems.*

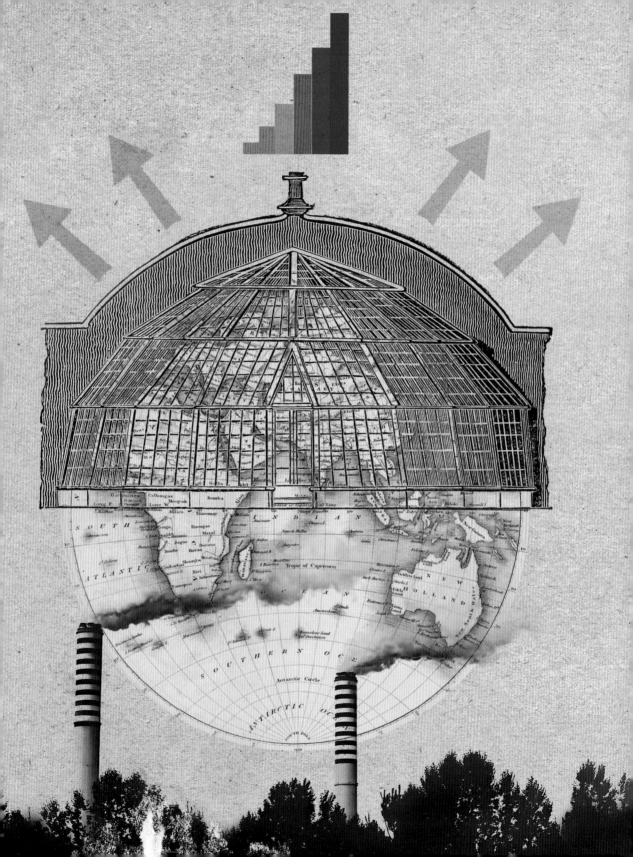

**February 19, 1859**
Born in Vik, Sweden

**1876**
Enters the University of Uppsala to study chemistry, physics, and mathematics

**1884**
Receives Doctorate from the Swedish Academy of Sciences with thesis on electrolytes

**1884**
Secures research post at the University of Uppsala

**1891**
Appointed to a position lecturing in physics at the University of Stockholm

**1895**
Appointed professor of physics at the University of Stockholm

**1896**
Publishes paper "On the Influence of Carbonic Acid in the Air Upon the Temperature of the Ground," which highlights the importance of carbon dioxide as a warming gas

**1900**
Publishes his book *Lärobok i Teoretisk Elektrokemi* (Textbook of Theoretical Electrochemistry)

**1903**
Awarded the Nobel Prize in Chemistry

**1905**
Becomes rector of the Nobel Institute for Physical Research at Stockholm

**1906**
Publishes his book *Världarnas Utveckling* (Worlds in the Making), which includes a calculation of how cold the Earth would be without the greenhouse effect

**October 2, 1927**
Dies in Stockholm, Sweden

# SVANTE ARRHENIUS

**Svante Arrhenius was a Swedish** Nobel Prize-winning chemist who first recognized the significance of atmospheric carbon dioxide in the greenhouse effect.

Born in Vik, Sweden, Arrhenius excelled at mathematics and physics in school. He was better known in his own lifetime for his contribution to chemistry than for quantifying the role of carbon dioxide. With his early research into the way certain chemical compounds, called electrolytes or salts, behave in solution, Arrhenius pioneered the scientific discipline we now call physical chemistry. In conjunction with Russian-German chemist Wilhelm Ostwald he developed the idea that dissolved molecules of these electrolytic compounds dissociate into charged atoms known as ions that carry electrical charges. This was the major insight that secured Arrhenius the Nobel Prize for Chemistry in 1903.

Arrhenius was aware of the work of British chemist John Tyndall who had already shown that carbon dioxide and water vapor could absorb heat in the laboratory although Tyndall had focused mainly on water vapor, which plays a bigger role in warming the atmosphere.

Arrhenius identified changing carbon dioxide levels as an important source of warming. He used data from astronomers who had been measuring the temperature of the Moon's surface with a new instrument—known as a bolometer—that could detect infrared light radiated by the Moon. In a scientific paper published in 1896, Arrhenius was able to calculate how much of this infrared radiation was lost as it catalyzed through the Earth's atmosphere and he was able to link this loss to absorption by carbon dioxide. He also calculated just how much warmer the Earth would become due to an increase in the amount of carbon dioxide in the atmosphere and formulated a mathematical relationship between changes in carbon dioxide levels and changes in global temperature.

Arrhenius used the newly developed physics of blackbody radiation to show that the theoretical temperature of the Earth, given its distance from the Sun, would be around 6.8°F if heat were not being trapped within the atmosphere—that is, the Earth would be a frozen ice planet without the protective warming blanket provided by the greenhouse effect. He popularized the role of carbon dioxide in warming our planet in a book published in 1906.

Despite his achievements in chemistry, we now acknowledge Svante Arrhenius as the scientist who first calculated the size of the greenhouse effect due to increasing amounts of carbon dioxide in the atmosphere.

*Leon Clifford*

# ACID RAIN & ATMOSPHERIC POLLUTION

## the 30-second meteorology

**3-SECOND BREEZE**
Pollution adds to the
atmospheric concentration
of certain gases that react
with moisture in the air to
make acids, which can lead
to acid rain.

**3-MINUTE SHOWER**
Air pollution is not new.
It has been a problem in
busy cities since Roman
times and was well known
to Londoners in 19th-
century England. Since the
beginning of the Industrial
Revolution, ever-greater
quantities of sulfur dioxide
have been emitted into
the air causing acid rain.
Research now suggests
that worsening air pollution
owing to rapid, widespread
industrialization, in Asia
particularly, can even
affect the weather.

Pure water is neither acidic nor alkaline but the water that falls as rain contains impurities that can make it acidic, and when this happens the result is known as acid rain. Naturally occurring carbon dioxide ($CO_2$) in our atmosphere reacts with water to produce carbonic acid ($H_2CO_3$) and this means rainfall is naturally slightly acidic. But other impurities due to pollution as well as natural events such as volcanic eruptions and lightning can lead to precipitation being even more acidic than normal. Huge energies in lightning flashes cause chemical reactions that combine nitrogen and oxygen in the atmosphere to create nitrogen dioxide ($NO_2$), which reacts with water to make nitrous acid ($HNO_2$) and the highly corrosive nitric acid ($HNO_3$). Volcanic eruptions can inject sulfur dioxide ($SO_2$) into the atmosphere, which reacts with oxygen and water vapor to produce the equally corrosive sulfuric acid ($H_2SO_4$). In addition to these natural causes, atmospheric pollution from human activities is contributing additional carbon dioxide, nitrogen dioxide and sulfur dioxide into our atmosphere and so increasing the acidity of rainfall. Vehicle exhaust emissions, oil refining, and coal-fired power stations emitting sulfur dioxide pollution are the main contributors to human-induced acid rain.

**RELATED TOPICS**
See also
CLOUDS
page 22

RAIN
page 24

SNOW
page 28

**3-SECOND BIOGRAPHIES**
JOHN EVELYN
1620–1706
English diarist who noted the impact of corrosion on ancient Greek statues carved of marble

ROBERT ANGUS SMITH
1817–84
Scottish chemist who pioneered research into air pollution and coined the term "acid rain" in 1872

JAMES PITTS
1921–2014
American smog researcher whose work contributed to California's clean-air laws

**30-SECOND TEXT**
Leon Clifford

*Sulfate and nitrate emissions cause acid rain, which pollutes the air, soil, and water.*

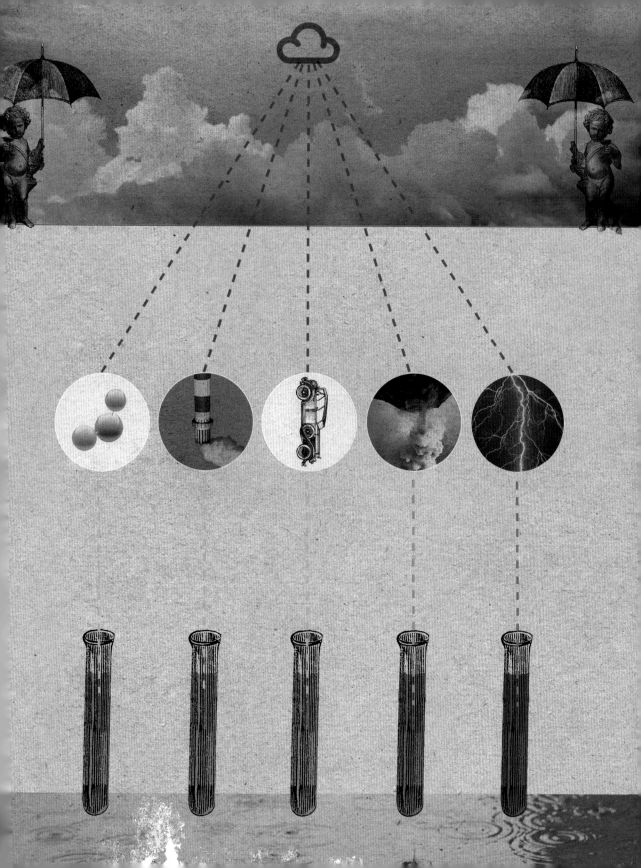

# CONTRAILS

## the 30-second meteorology

## Contrails are the white streaks

that appear in the wake of high-flying aircraft. Like clouds they form from the condensation of water vapor in the atmosphere and consist of billions of tiny water droplets and, more usually, ice crystals that are suspended in the air. Unlike clouds contrails are artificial. They tend to appear at heights of 5 miles and more where the air is cold and humid but they can form at lower altitudes if the air is cold enough. Contrails may form in the exhaust stream of aircraft engines, along the edges and surfaces of wings or at wing tips. Several processes are involved. Aircraft wings change the pressure of the air that they fly through and this change in pressure can cause atmospheric water vapor to condense. Aircraft engines inject both particles and exhaust gases including water vapor into the air. The exhaust particles can act as seeds for water condensation. Also the water vapor in the exhaust increases the humidity of the air in the exhaust stream and this can trigger further condensation. Exhaust water vapor takes time to cool, which explains why there sometimes appears to be a gap between the aircraft and the contrails that follow it.

**RELATED TOPICS**
See also
LAYERS OF THE ATMOSPHERE
page 18

CLOUDS
page 22

**3-SECOND BIOGRAPHY**
HERBERT APPLEMAN
1917–2013
American meteorologist who pioneered contrail research, plotting the temperature and pressure combinations that would give rise to contrails

**30-SECOND TEXT**
Leon Clifford

**3-SECOND BREEZE**
Aircraft create artificial clouds called contrails caused by water vapor condensation triggered by engine exhaust and by changes in pressure around the aircraft.

**3-MINUTE SHOWER**
How do contrails affect the weather? Contrails are essentially clouds and, like clouds, they reflect sunlight, which acts to cool the Earth and absorb heat radiated from the ground, which helps keep us warm. On balance, high clouds are thought to warm the Earth. However, when U.S. airspace was closed after the attacks of 11 September 2001 and U.S. skies were clear of contrails there was a small but measurable increase in temperatures.

*Contrails, a contraction of "condensation trails," are humanmade clouds that mark the path of jets flying in the upper troposphere.*

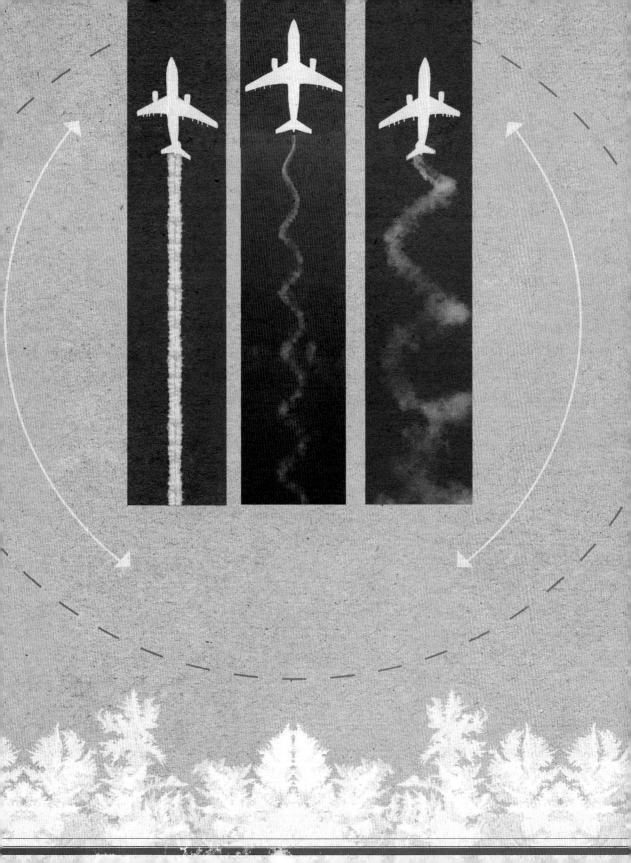

# WEATHER CYCLES

**electromagnetic radiation** A form of energy that is described in terms of vibrations in electric and magnetic fields. Visible and ultraviolet light (shortwave radiation), infrared radiation (heat or longwave radiation), x-rays, and gamma rays, microwaves and radiowaves are all examples of electromagnetic radiation. Most of the energy emitted by the Sun is in the form of electromagnetic radiation. It travels at the speed of light and transports solar energy from the Sun to the Earth. The different types of electromagnetic radiation are distinguished by their wavelength (or color) and by their frequency; the shorter the wavelength the higher the frequency and vice versa. The smallest unit of electromagnetic radiation is the photon. Photons of light with shorter wavelengths, and higher frequencies, carry greater energy than photons of longer wavelength light. Different atmospheric gases transmit and absorb electromagnetic radiation of various wavelengths in different ways. Carbon dioxide is transparent to visible light but absorbs infrared and ozone absorbs ultraviolet light but transmits visible light. When an atmospheric gas absorbs electromagnetic radiation, it acquires the energy in that radiation and so warms.

**El Niño-Southern Oscillation (ENSO) cycle** A pattern of warming and cooling of the temperature of the sea surface in the tropical Pacific Ocean which is linked to changes in the atmosphere above the ocean. The warming part of the cycle is called El Niño and the cooling part is called La Niña and the related fluctuations in the pattern of air pressure over the tropical Pacific are referred to as the Southern Oscillation. This linked combination of sea-surface temperature variations and atmospheric pressure changes is known as the ENSO cycle.

**interglacials** A period of warmer than average global temperatures within an ice age is known as an interglacial. During interglacials the icesheets retreat before advancing again during a new period of glaciation. There have been at least five major ice ages in the history of the Earth. The Earth is currently experiencing an ice age known as the Quaternary which began around 2.58 million years ago. It has been characterized by a series of many glaciations when giant icesheets covered much of the planet. These have been interspersed by warmer interglacial periods when the icesheets retreated. Our current ice age has involved many cycles of glaciation separated by interglacials. Typically, glacial periods last between 40,000 and 100,000 years and interglacial periods last around 10,000 years. We are living during an interglacial when the icesheets have retreated to Greenland and Antarctica; it has already lasted more than 10,000 years.

**megadroughts** A period with extremely low precipitation that lasts for more than two decades is known as a megadrought. In the past, megadroughts occurred in various parts of the world, according to historical climate data. The term refers to the duration of the drought rather than the intensity or severity. Many scientists believe megadroughts will become more common as the world warms due to climate change. NASA has warned that parts of America's southwest could experience a megadrought this century.

**oscillations** Cyclical patterns in the weather and in the climate system are sometimes called oscillations. They can take place over periods as short as weeks and months or over periods of decades or even longer. They can involve any aspect of the weather—rainfall, pressure, and ocean temperature—and, typically, they will link and connect several different features together. The Madden-Julian Oscillation, a fluctuation in tropical weather that drives a pulse of cloud and rain along the equator every 30–60 days, is one example. Another is the Pacific Decadal Oscillation that results in alternate warming and cooling of subsurface waters in the Pacific Ocean over a period of around 30 years. The cycle of glacial and interglacial periods during ice ages is also a form of oscillation that takes thousands of years.

**solar energy** Solar energy powers the weather and is the ultimate source of heat in the climate system. In meteorology, solar energy is that fraction of the Sun's energy output that is received by the Earth. Solar energy is generated within the Sun through the process of nuclear fusion and reaches the Earth in the form of light and other electromagnetic radiation. X-rays and ultraviolet light emitted by the Sun are absorbed by the atmosphere. The remaining solar energy reaching the Earth is either absorbed by the surface or reflected back into space by clouds and ice. The term solar energy is also applied to the electricity or heat generated by devices that capture this energy from the Sun.

# THE MADDEN-JULIAN OSCILLATION (MJO)

## the 30-second meteorology

The Madden-Julian Oscillation is hard to define, and even harder to understand, but we know it when we see it. Spanning several thousand miles the MJO is an eastward-moving pattern of cloud and rainfall near the equator that often recurs every 30–90 days. It is the tropics' closest equivalent to the weather patterns that sweep across the mid-latitudes every week or so, and is the most significant contributor to tropical weather on weekly to monthly timescales. The tropics don't have weather in the same way as the mid-latitudes do. By and large they are uneventful regions, save for the seasonal monsoons and the occasional hurricane. Indeed, over large swathes of the tropics the prevailing winds are so light the regions are called the doldrums. The MJO is something of an exception. It generally begins in the western Indian Ocean, moves eastward at speeds of 4–8 yd/s before dying out in the colder eastern tropical Pacific. The pattern consists of a wet, rainy region of ascending air flanked by drier regions on either side. But here's the rub: we don't fully understand what's going on. The phenomenon has wave-like attributes but in many ways it is more like a translating pattern, and we still don't know what determines its propagation speed or time and space scales.

*Rain, waves, convection, and balance—they all come together to produce the MJO. Quite how they do it no one knows, and that is the allure of meteorology.*

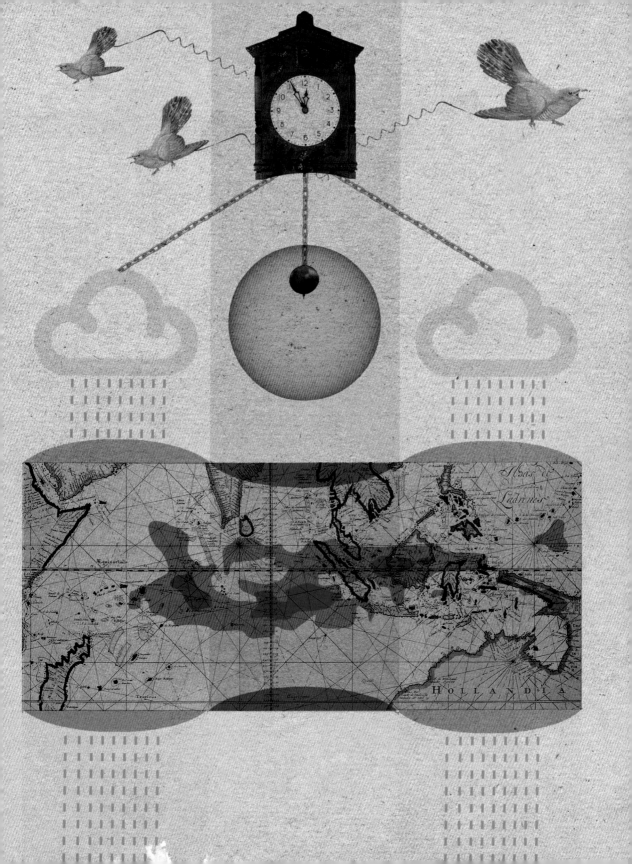

# EL NIÑO & LA NIÑA

## the 30-second meteorology

Every few years enough warm water builds up on the equator in the Pacific Ocean to trigger one of the most dramatic natural climate fluctuations in the Earth's climate. It warms the sea surface and the air above and extends to the coast of the Americas. This sea-surface warming event is known as an El Niño, and its flipside, called La Niña, can cool the surface of the Pacific Ocean and absorb heat from the atmosphere. Global surface temperatures are measurably warmer during years following El Niño and cooler following La Niña. Both events are part of a natural cyclical system of heat exchange between the Pacific Ocean and the atmosphere. They are coupled to fluctuations in atmospheric pressure over the Pacific Ocean, called the Southern Oscillation, that moves in step with the changes in sea-surface temperature. This combination of periodic ocean temperature variations and related atmospheric pressure changes is known as the El Niño-Southern Oscillation (ENSO). The full ENSO cycle typically occurs every few years and affects wind and rainfall patterns around the globe. It has been linked to droughts in Africa and cold winters as far distant as northern Europe.

**3-SECOND BREEZE**
El Niño and La Niña are part of a cyclical pattern of heat exchange between the Pacific Ocean and the atmosphere affecting weather around the world.

**3-MINUTE SHOWER**
A major El Niño event took place in 1997 and 1998. It warmed the Earth so much that it drove up average global surface temperatures to make 1998 the hottest ever recorded. Only years later did global warming and subsequent El Niño events break the record again. Some climate scientists believe that the Pacific Ocean is storing a significant amount of heat and that, in the near future, this heat may be released to accelerate global warming.

**RELATED TOPICS**
See also
TRADE WINDS
page 62

GILBERT T. WALKER
page 84

PACIFIC DECADAL
OSCILLATION
page 130

**3-SECOND BIOGRAPHY**
JACOB BJERKNES
1897–1975
Norwegian-American meteorologist who explained the physics of the ENSO phenomenon in 1969

**30-SECOND TEXT**
Leon Clifford

*The name El Niño, Spanish for the boy or Christ child, was used because the event was observed to occur around Christmas time.*

# NORTH ATLANTIC OSCILLATION (NAO)

## the 30-second meteorology

### 3-SECOND BREEZE
For eastern North America and Europe, the NAO is the single most important factor in year-to-year differences in the weather.

### 3-MINUTE SHOWER
Due to its irregular nature, it has been suggested by some meteorologists that a better name for the NAO would be "Not An Oscillation" and computer model studies previously suggested that the NAO was fundamentally unpredictable. However, the latest research shows that this was overly pessimistic and we now know that skillful predictions of the winter NAO can often be made up to a few months ahead.

**Dramatic swings in weather can** occur from one winter to the next in the middle latitudes but these seemingly complicated changes often form a simple weather pattern across the North Atlantic. The largest changes in pressure occur near Iceland and the Azores. A kind of weather seesaw means that when the Icelandic pressure is lower (higher) than usual, the Azores pressure is often higher (lower) than usual. This seesaw in pressure is the North Atlantic Oscillation—NAO for short. A change in the NAO heralds changes in all aspects of weather across eastern USA, Canada, and Europe. For example, the NAO was strongly positive in winter 1999/2000 and temperatures in northern Europe were mild but fierce storms plowed into France and Germany, taking lives and causing billions of euros worth of damage. In contrast, the NAO was extremely negative in winter 2009/10 and northern Europe was still and dry but went into deep freeze for months. Despite 150 years of reliable observations, little pattern can be found in the record of the NAO as it is too irregular to be a genuine oscillation. However, a variety of factors drives the NAO, ranging from El Niño, way off in the remote Pacific, to changes in the sunspot cycle above and the Atlantic Ocean below.

### RELATED TOPICS
See also
JET STREAMS
page 50

BLOCKING, HEATWAVES
& COLD SNAPS
page 58

GILBERT T. WALKER
page 84

### 30-SECOND TEXT
Adam A. Scaife

*When the winter NAO is in its positive phase, the Atlantic jet stream is strong and it brings mild, wet, and stormy weather to eastern USA and northern Europe while eastern Canada and southern Europe are left in the cold. However, this all reverses when the NAO is in its negative phase, as it was in winter 2009/10.*

# QUASI-BIENNIAL OSCILLATION (QBO)
## the 30-second meteorology

Every 14 months or so, the winds blowing around the equator at high altitude reverse direction and flow the opposite way. Some 14 months later they flip back again, taking about 28 months to complete the full cycle. This remarkable Quasi-Biennial Oscillation (QBO) was discovered by meteorologists following the frequent launching of weather balloons in the late 1950s; great consternation followed as to its cause. Several clues emerged but it took nearly two decades before American scientists Richard Lindzen and James Holton showed that the QBO winds are driven by small-scale waves emanating from intense tropical weather systems below. These waves "break" like waves break on a beach and provide a systematic push to the wind. In fact, the QBO had been unwittingly seen nearly a century before. The plume from the devastating eruption in 1883 of Krakatau (an island volcano in the Indonesian archipelago) was tracked moving across the tropics at a speed that we now know matches the QBO winds. Although it may seem remote, the QBO is relevant to the Atlantic jet stream, which tends to strengthen when the QBO blows from the west, and weaken when the QBO blows from the east.

**RELATED TOPICS**
See also
JET STREAMS
page 50

ATMOSPHERIC WAVES
page 54

**3-SECOND BIOGRAPHIES**
ROBERT EBDON
1928–
Codiscoverer of the QBO in early weather balloon data who also carried out early work on the effect of the QBO on Atlantic meteorology

JAMES HOLTON
1938–2004

& RICHARD LINDZEN
1940–
American meteorologists who first demonstrated the counterintuitive mechanism behind the QBO

**30-SECOND TEXT**
Adam A. Scaife

**3-SECOND BREEZE**
Apart from the annual march of the seasons, the Quasi-Biennial Oscillation is the most regular slow variation in the atmosphere.

**3-MINUTE SHOWER**
Computer models based on the laws of physics can now simulate the QBO. These same computer models are also used to make our daily weather forecasts. So although it may seem like a remote curiosity, the striking regularity of the QBO and its effect on the Atlantic jet stream, storms, and extreme winter weather offer hope of improving very long-range weather forecasts.

*High-altitude winds circle the globe over the equator and completely reverse direction from westerly to easterly every 14 months.*

# PACIFIC DECADAL OSCILLATION (PDO)

## the 30-second meteorology

**The Pacific Decadal Oscillation is** the leading pattern of warm and cool phases in Pacific sea-surface temperature from one decade to the next, excluding general warming due to global average temperature changes. Changes in the PDO remain something of a mystery but they appear to result from a combination of responses to the tropical El Niño–Southern Oscillation (ENSO) and deepening or filling of the atmospheric low-pressure region near the Aleutian Islands. Although broadly similar in pattern to the ENSO, the PDO is much more active in the extratropical North Pacific and less active in the far eastern tropical Pacific. Such decadal variability of the oceans extends to the South Pacific and, to a small extent, into the Indian and Atlantic Oceans. This near-global pattern is called the Interdecadal Pacific Oscillation (IPO). Evidence from climate models and observations indicates that the PDO and IPO influence global average temperature. Their cold phases tend to increase the mixing of colder, deep Pacific Ocean waters with surface waters enough to temporarily reduce the rate of global warming caused by increasing greenhouse gases, an effect that occurred in the early part of this century.

### RELATED TOPICS
See also
GLOBAL WARMING &
THE GREENHOUSE EFFECT
page 110

EL NIÑO & LA NIÑA
page 124

### 3-SECOND BIOGRAPHY
SIR CHARLES WYVILLE
THOMSON
1830–82
Scottish zoologist and Chief
Scientist of the *Challenger*
expedition (1872–76) that made
the first major study of the
physical characteristics and
biology of the Pacific

### 30-SECOND TEXT
Chris K. Folland

*Strong positive (top) and negative (bottom) phases of the IPO have farreaching global consequences, resulting in feasts in some regions, famine in others. The maps show differences of ocean-surface temperature from a longterm average.*

### 3-SECOND BREEZE
The PDO and IPO are patterns of climate variability in the Pacific that were discovered by American, British, and Australian scientists in the late 1990s.

### 3-MINUTE SHOWER
The warm phase of the PDO causes periods of bountiful salmon in southern Alaska, whereas cold phases contribute to multiannual droughts in southwestern USA. Climate modeling studies suggest that "megadroughts" in this region centuries and millennia ago were associated with prolonged cold phases of the PDO and IPO. However, the cold phases increase rainfall over eastern Australia, aiding agriculture and water resources.

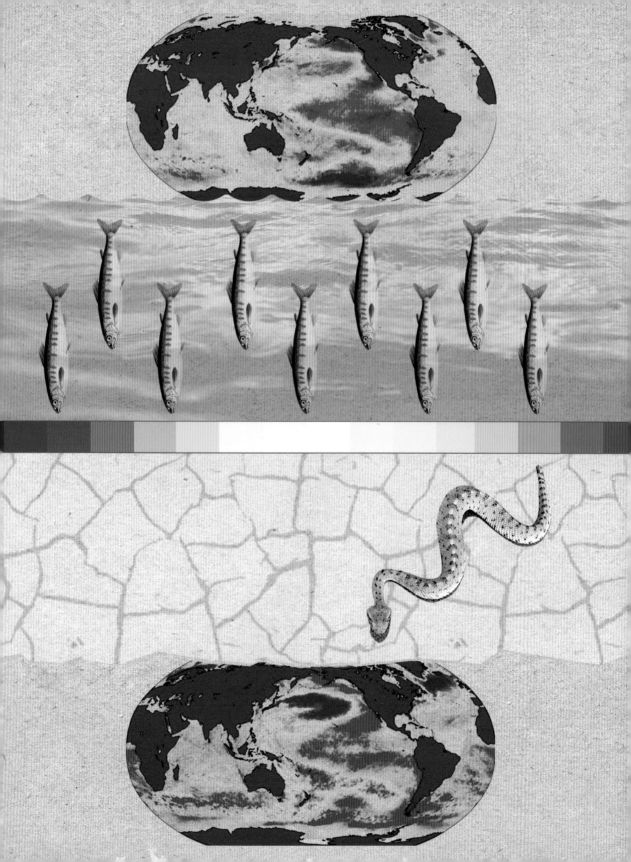

# ATLANTIC MULTIDECADAL OSCILLATION (AMO)

## the 30-second meteorology

### Climate in the North Atlantic

appears to go through cycles. Records show that alongside a long-term warming trend, sea-surface temperatures across much of the North Atlantic were warm between about 1925 and 1965 and then after 1995, but cool between 1965 and 1995. This cycle has been named the Atlantic Multidecadal Oscillation (AMO) and its phases have been linked to changes in climate across the world. In much of North America, for example, warm phases favor reduced summer rainfall, and the warm AMO phase contributed to the 1930s "Dust Bowl" drought. AMO influences also affect rainfall in Africa's Sahel region, northeast Brazil, Europe in summer, the Arctic climate, and even the Indian monsoon. Evidence from European and North American tree rings shows that the AMO has existed for many centuries, as the tree rings show the AMO influence on summer temperatures in these regions. The cause of the AMO has been investigated using climate models, which suggest changes in the movement of heat into the North Atlantic through speeding up and slowing down of the global-scale circulation of the ocean. An alternative hypothesis is that changes in the emissions of particles from pollution defined the sequence of recent AMO phases.

---

**3-SECOND BREEZE**
The AMO, an approximately 70-year cycle in North Atlantic Ocean temperature, has widespread effects on climate across much of the globe.

**3-MINUTE SHOWER**
The AMO affects the formation of hurricanes by modifying large scale weather systems in the subtropical Atlantic. As a result, the 1970s and 1980s had relatively few storms, while the decade after 1995 was extremely active. The most active season recorded was 2005, with 15 hurricanes including four that reached the strongest category 5. Hurricane Katrina became infamous for causing flooding in New Orleans, which claimed over 1,800 lives.

---

**RELATED TOPICS**
See also
MONSOONS
page 66

NORTH ATLANTIC
OSCILLATION
page 126

PACIFIC DECADAL
OSCILLATION
page 130

HURRICANES & TYPHOONS
page 146

**3-SECOND BIOGRAPHY**
JACOB BJERKNES
1897–1975
Norwegian-American meteorologist who first noted the relative warmth of the North Atlantic in the period between the 1930s and 1960s, and proposed that it was related to changes in the movement of oceanic heat

**30-SECOND TEXT**
Jeff Knight

*The alternating warm and cool phases of the AMO have influenced climate patterns for at least several centuries.*

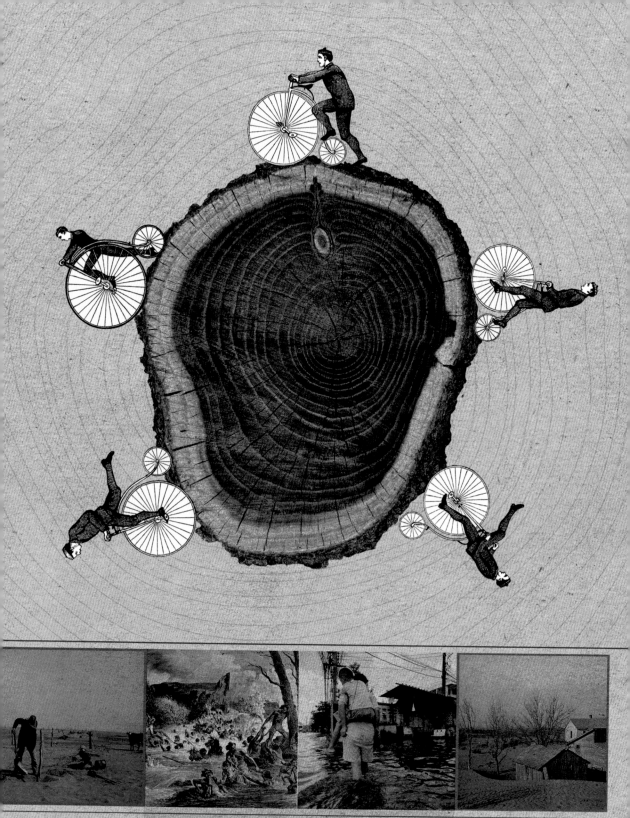

# PAST CLIMATES & THE LITTLE ICE AGE

## the 30-second meteorology

### 3-SECOND BREEZE
Our planet's climate history is long and varied with periods in the last millennium when climate differed substantially from the modern day.

### 3-MINUTE SHOWER
Cold winters in Europe during the Little Ice Age left a cultural impression recorded in paintings by Bruegel and novels by Dickens. Elsewhere in the world, measures like tree rings are used to reconstruct climate. Nevertheless, gauging how widespread the Little Ice Age was remains difficult. Both low solar activity and a sequence of volcanic eruptions are thought to have cooled climate during this era.

It is possible that in the Earth's distant past the global climate was so cold that icesheets reached the equator. While evidence for this is debated, geology provides ample clues of a long and complex history of the Earth's climate variations. Drivers of these fluctuations include continental drift, which repositions the continents in latitude and changes the paths of ocean currents, episodes of increased volcanic outgassing, and mountain-building phases that alter atmospheric composition through accelerated weathering. These changes are on such a vast scale that it is hardly surprising that climate could have been very different from today. In the last few million years, when the planet was geologically similar to the present, climate has fluctuated between ice ages and interglacials as a result of cycles in the Earth's rotation and orbit. These influences continued into the present interglacial, leading to a peak in global temperatures between 9,000 and 5,000 years ago. Since then, global climate has generally cooled, until the current period of human-influenced warming. In the last millennium, there have been shorter-term fluctuations, including the Medieval Warm Period (ca. 950–1250) and the Little Ice Age (ca. 1500–1850). Doubt surrounds whether these variations were truly global, however, rather than just regional events.

**RELATED TOPICS**
See also
SUNSPOTS & CLIMATE
page 82

WEATHER RECORDS
page 92

GLOBAL WARMING &
THE GREENHOUSE EFFECT
page 110

MILANKOVITCH CYCLES
page 138

**3-SECOND BIOGRAPHY**
HUBERT LAMB
1913–97
British climatologist who was among the first to study climate variability in the last millennium

**30-SECOND TEXT**
Jeff Knight

*The Earth's climate has varied dramatically throughout geological time and influenced historical events in all parts of the world.*

**May 28, 1879**
Born in Dalj in what is now Croatia and was then part of the Austro-Hungarian Empire

**1896–1902**
Studies Civil Engineering at the Technische Universität Wien (Vienna University of Technology)

**1903**
Military service

**1904**
Awarded doctorate for a thesis on the use of building material such as concrete in curved structures

**1905–12**
Works as a civil engineer on a number of large construction projects

**1909**
Takes up the Chair of Applied Mathematics at Belgrade University

**1912**
Publishes the first of a series of papers linking the Sun, the Earth's orbit, and climate

**1914**
Interned for being an ethnic Serb following Serbia's secession from Austria-Hungary

**1919**
Returns to Belgrade University to work as a professor

**1920**
Publishes his ideas on climate and orbit in a book

**1938**
Publishes a paper explaining the mathematical connection between insolation (incoming solar radiation) and the position of snow lines and the edges of icesheets

**1941**
Completes a work that collates all his research and ideas on insolation, weather, climate and ice ages

**December 12, 1958**
Dies in Belgrade in what is now Serbia and was then part of Yugoslavia

**1970**
Crater on the Moon is named after Milankovitch

**1976**
Major paper in *Science* confirms link between Earth's orbit and ice ages

# MILUTIN MILANKOVITCH

## Milutin Milankovitch, an ethnic

Serb, was born in the Balkans in 1879 within the territory of what was then the Austro-Hungarian Empire. Mathematical talent showed from an early age and he began his professional life as a civil engineer, helping to build bridges, dams and hydro-electric power stations and developing a particular expertise in the properties and uses of concrete. His focus gradually shifted from civil engineering works toward meteorology and climate following his appointment in 1909, at the age of 30, to an academic role in applied mathematics at Belgrade University.

At Belgrade, Milankovitch became intrigued by the mystery of the repeating ice ages that had been discovered in the geological record. He thought that there was some kind of connection between the Sun and these cyclical ice ages and that this connection involved the orbit of the Earth. Realizing that rigorous mathematical methods had not been applied to this meteorological and climatic mystery, he went about analyzing how the Earth's orbit and the tilt of the planet's axis changed over time.

This belief that mathematics could explain climate phenomena led Milankovitch to piece together a relationship between the Earth's orbit and the observed pattern of repeating ice ages. In 1912 he suggested that there were longterm cycles at work in the climate that had their origins in the astronomical motion of the Earth—what we now call Milankovitch Cycles.

As a Serb, Milankovitch was interned on the eve of World War I when Serbia broke away from Austria-Hungary. The intercession of a former university tutor led to his release and he was allowed to work in Belgrade for the remainder of the conflict. After the war, he returned to Belgrade University where he continued to develop his ideas, becoming an internationally recognized academic and the author of many books.

In the years following Milankovitch's death in 1958, his astronomical explanation for ice ages fell out of favor. But his ideas were gradually rehabilitated and his theory was essentially vindicated in 1976 with the publication of a major scientific paper in the prestigious journal *Science*. This found evidence of longterm climate cycles in seafloor sediments and its conclusion, "changes in the earth's orbital geometry are the fundamental cause of the succession of Quaternary ice ages," proved that Milankovitch had been right.

*Leon Clifford*

# MILANKOVITCH CYCLES

## the 30-second meteorology

**The apparently unchanging** yearly cycle of the seasons is one of the most reliable features of our weather. In fact, the seasons slowly change over millennia through variations in the Earth's rotation and orbit caused by the gravitational pull of the Moon and planets. The effect is to shift the date of the Earth's annual closest approach to the Sun. It takes 22,000 years to complete a cycle, so although the Earth is now closest to the Sun on January 3 each year, 11,000 years ago this occurred in July. Also, the Earth's orbit is slightly noncircular—currently the difference between the closest and farthest points from the Sun is 3 million miles. Over about 100,000 years this difference cycles between almost zero and about 9 million miles. Finally, the tilt of the Earth's axis varies between 22.1 and 24.5 degrees over about 41,000 years. These three cycles do not change the total energy the Earth receives from the Sun, but they do modify the amount in each season and hemisphere. Astronomer Milutin Milankovitch suggested this variation was the initiator of the ice ages. Additional amplifying processes are necessary, however, to explain how these small changes caused the polar icesheets to advance over much more of the northern hemisphere than they do today.

**3-SECOND BREEZE**
Very long astronomical cycles change the distribution of solar energy and are thought to be the "pacemakers" of ice ages.

**3-MINUTE SHOWER**
Evidence from cores drilled from Greenland and Antarctica icesheets, and from land and ocean sediments, allows the history of the ice ages to be reconstructed going back millions of years, showing 41,000- and 100,000-year Milankovitch cycles. The most recent million years featured cycles of ice ages and interglacials approximately every 100,000 years and the last ice age ended about 11,000 years ago.

**RELATED TOPICS**
See also
SEASONS
page 20

PAST CLIMATES
& THE LITTLE ICE AGE
page 134

MILUTIN MILANKOVITCH
page 136

**3-SECOND BIOGRAPHY**
LOUIS AGASSIZ
1807–73
Swiss-American geologist who was the first to propose that the Earth had experienced an ice age

**30-SECOND TEXT**
Jeff Knight

*A century after Louis Agassiz's discovery that the Earth had experienced a succession of ice ages and interglacials, Milutin Milankovitch provided the explanation based on cycles in the Earth's rotation and orbit.*

# EXTREME WEATHER

**Fujita scale** The intensity of tornadoes is measured in terms of the Fujita scale. Invented by Tetsuya Theodore Fujita and Allen Pearson in 1971, it categorizes tornadoes in terms of the damage that they do and an estimate of associated wind speeds. The scale originally ranged from F0, which causes light damage such as broken branches on trees and damage to signboards on buildings, up to F5, which can sweep away strongframed houses and hurl cars or vehicle-sized missiles in excess of 100 yards through the air. An enhanced Fujita scale was introduced in the USA in 2007 that provides more accurate estimates for the wind speed of the damaging gusts that cause most havoc.

**graupel** A form of precipitation with the texture of small soft ice pellets, sometimes called soft hail. However, graupel is not hail and it is not sleet; it has a different texture and structure and is formed in a different way. It results when droplets of supercooled water in suspension in the atmosphere coalesce and freeze around a falling snowflake. It can be triggered by a band of rain moving into a chilled air mass.

**oscillations** Cyclical patterns in the weather and the climate system are sometimes called oscillations. They can occur over weeks or months, over decades, or even longer. They can involve any aspect of the weather—rainfall, pressure, and ocean temperature—and, typically, they will link several different features together. The Madden-Julian Oscillation, a fluctuation in tropical weather that drives a pulse of cloud and rain around the equator every 30–60 days, is one example. Another, the Pacific Decadal Oscillation, results in alternate warming and cooling of subsurface waters in the Pacific Ocean over a period of around 30 years. The cycle of glacial and interglacial periods during ice ages is also a form of oscillation that takes thousands of years.

**plasma** When atoms are stripped of one or more of their negatively charged electrons they become positively charged and are referred to as ions and can form a plasma. A plasma is an ionized gas consisting of freemoving electrons and their parent atoms which is electrically neutral in its entirety. However, a plasma, unlike an un-ionized gas, can conduct an electric current. Plasmas tend to be unstable and shortlived unless some mechanism maintains them. The discharge path of lightning through the atmosphere consists of hot air plasma. Plasma is also found in the upper portion of our atmosphere known as the ionosphere where incoming solar radiation maintains a plasma by continuously knocking electrons from atoms of oxygen and other gases.

**stratosphere** The layer of the Earth's atmosphere between an altitude of around 7 and 30 miles above sea level. The stratosphere begins closer to the surface at the poles (around 4 miles) and higher above the surface at the equator (around 11 miles). It features extremely cold, thin, dry air and is home to the ozone layer that protects us from much of the Sun's damaging ultraviolet light. Unlike the lower atmosphere, air temperature in the stratosphere increases with altitude due to the warming effect of this ozone, which is heated by the energy from the ultraviolet light it absorbs.

**supercell storm** A rare type of storm that is formed around a rotating updraft of air and that can cause extreme weather conditions on the ground. Supercells have been associated with violent tornadoes, dangerous lightning, baseball-size hail, strong winds, and heavy downbursts of rain leading to flash flooding. They are believed to be caused when wind shear tilts a horizontal vortex (or rotating air mass) so that it rotates around a vertical axis, creating a powerful updraft known as a mesocyclone. This rotational element distinguishes a supercell storm from more common singlecell and multicell storms.

**supercooling/supercooled water** Supercooling occurs when a liquid is cooled below its normal freezing point but does not solidify. Supercooled water droplets are found in high-altitude clouds where the temperature of the air is below the freezing point of water. This supercooled state can only be achieved in droplets that do not contain impurities or aerosols that would otherwise act as seeds to trigger crystallization. Research suggests that the phenomenon of supercooling may be due to the molecules of water arranging themselves in a way that is incompatible with crystallization.

**temperature inversion** In the troposphere (the lowest layer of the Earth's atmosphere) air temperature usually falls with increasing altitude but sometimes it can increase, resulting in a blanket of warm air sitting above a layer of cooler air. This is known as a temperature inversion. Rain falling through a temperature inversion can freeze, causing freezing rain. If the air below the inversion is sufficiently humid fog can form. Over populated areas, temperature inversions can act as a lid that traps pollution near the ground.

**vortex/vortices** A rotating mass of fluid is known as a vortex. In meteorology, a vortex usually refers to a rotating mass of air. This rotation can take place around a low-pressure system, as in a hurricane or typhoon.

# THUNDERSTORMS & LIGHTNING

## the 30-second meteorology

**Lightning is mainly observed from** tall, ice-bearing cumulonimbus clouds and it is generally agreed that the thermoelectric effect of ice is important in the formation of the electric charge separation that precedes it. This involves the temperature-dependent tendency for water molecules to dissociate into negative and positive ions, and results in negative and positive charges at the warm and cold ends respectively of a piece of ice. When graupel (soft hail) hits a supercooled water droplet it joins and freezes, the whole remaining warmer than its environment because of latent heat release. However, when graupel collides with an ice particle they instantaneously form a single piece of ice of nonuniform temperature before generally bouncing apart. Having been thermoelectrically positively charged by this brief encounter, the colder ice particle is carried by updrafts to the cloud's higher reaches while the heavier, negatively charged, graupel falls earthward, often melting. Air is a good insulator, allowing the buildup of enormous electrical charges in this way, but these eventually overcome resistance and release a lightning stroke, along which a massively powerful electric current travels. Air molecules are split as they are heated to 36,000–54,000°F, resulting in a brightly glowing, rapidly expanding plasma and a shockwave that generates a thunderclap.

# HURRICANES & TYPHOONS

## the 30-second meteorology

### These intense storms are Tropical

Cyclones (TCs) with top surface wind speeds over 70 mph. In the Northwest Pacific they are known as typhoons; in the Northeast Pacific and North Atlantic they are called hurricanes. TCs undergo fascinating transformations in their short life (a few weeks at most). For example, Atlantic hurricanes are born from westward-traveling eddies over Africa that can grow into coherent cyclonic vortices, depending on their strength and environmental factors. As the storm develops, a rotating column of air is confined by the air flowing around it, which permits the nurturing and shielding of towering hot, moisture-laden, and rainy clouds fed by the warm ocean. This process, combined with water temperature almost 80°F, can transform the moist vortex into a TC: a powerful engine extracting heat energy from the ocean. The mightiest TCs can dissipate more power than the total world generation of electricity in an area no bigger than Cuba. Mature TCs show often intriguing and stunning rainy spiral waves rotating around the hurricane eye in a few hours. In a matter of days, these bands can redistribute momentum within the TC and drive considerable TC intensity changes. TCs often terminate their lifecycle with a landfall or will curve away from the tropics; sometimes reintensifying, but eventually dissipating.

**3-SECOND BREEZE**
Originating as low-pressure systems over warm tropical waters, these circular storms can wreak havoc as they whip across oceans and land.

**3-MINUTE SHOWER**
In the past, a TC could kill thousands of people by surprise, especially from associated coastal flooding. Scientific studies suggest that TC intensity may be increasing due to warmer sea surface temperatures resulting from climate change. Numerical weather predictions can greatly mitigate the deadly impact of TCs—track forecasts are now available one week in advance—provided proper disaster risk-reduction actions are taken.

**RELATED TOPICS**
See also
WEATHER FORECASTING
page 98

CLIMATE PREDICTION
page 102

**3-SECOND BIOGRAPHY**
KERRY ANDREW EMANUEL
1955–
American meteorologist who has contributed to the understanding of mechanisms for TC intensification, lifecycle, and climatology

**30-SECOND TEXT**
Gilbert Brunet

*Hurricanes and typhoons are both examples of a vortex of air rapidly rotating around a core of low pressure—the "eye" of the storm.*

**May 23, 1917**
Born in West Hartford,
Connecticut

**1938**
Receives BA from
Dartmouth College,
New Hampshire

**1940**
Receives Masters from
Harvard

**1942–46**
Meteorologist in
U.S. Army Air Corps

**1948**
Receives Doctorate in
Meteorology from the
Massachusetts Institute
of Technology (MIT)

**1963**
Publishes the
foundations of chaos
theory in his paper
"Deterministic
Nonperiodic Flow"
in the *Journal of
Atmospheric Sciences*

**1969**
Awarded the Carl-Gustaf
Rossby Research Medal
by the American
Meteorological Society

**1973**
Receives Symons Gold
Medal of the Royal
Meteorological Society

**1983**
Receives Crafoord prize
of the Royal Swedish
Academy of Sciences

**1993**
Publishes his book
*The Essence of Chaos*

**1987–2008**
Becomes Professor
Emeritus at MIT, where
he remains for the rest
of his life

**1991**
Receives Kyoto Prize for
science for his discovery
of deterministic chaos

**2000**
Receives International
Meteorological Prize of
the World Meteorological
Organization

**2004**
Awarded the Lomonosov
Gold Medal by what is
now the Russian Academy
of Sciences and the Buys
Ballot medal by the Royal
Netherlands Academy of
Arts and Sciences

**April 16, 2008**
Dies aged 90

# EDWARD NORTON LORENZ

**Meteorologist and mathematician**
Edward Lorenz was a lifelong New Englander. He studied at Dartmouth College then at Harvard with George Birkhoff who had worked on "dynamical systems," which would be at the core of Ed's future work. World War II intervened, and Ed served as a meteorologist. After the war ,he completed a doctorate at MIT and in the 1950s took a visiting position at UCLA where he began a program of numerical forecasting using an early electronic computer and simple sets of coupled equations as approximations to the equations of atmospheric flow. Ed was well ahead of his time: many meteorologists were using linear statistical forecast methods of which he was skeptical.

During this work Ed made one of the greatest discoveries of his career. He had been running his computer model, which had set out the solution of his weather equations as 12 numbers (he later refined this to just 3 with help from a colleague). He then retyped the 12 numbers into the computer model and set it running again while he went for coffee. On his return he discovered that, despite starting from apparently the same conditions, the new solution was entirely different from the original. Ed realized that this was due to a tiny change he had introduced when reentering the 12 numbers. He had inadvertently demonstrated the existence of deterministic chaos, whereby the smallest of changes in the initial state can soon grow to produce wildly different outcomes. His later quote sums this up: "Two states differing by imperceptible amounts may eventually evolve into two considerably different states." This has since been termed the "butterfly effect" after Ed's papers on the sensitive dependence of the meteorological state on initial conditions—although he actually used the flapping of a seagull's wings in his original analogy.

Ed's work showed that relatively simple sets of equations could lead to complicated dynamics through a weird mathematical entity called a strange attractor. This remarkable concept is described by fractal geometry and lies at the heart of what we now call "chaos theory." His discoveries have reshaped modern weather forecasting—necessitating the production of multiple "ensembles" of forecasts to take into account the chaotic sensitivity to small errors.

Ed Lorenz's farreaching discoveries led to a radical shift in the way meteorologists, other scientists, and mathematicians understand the world. He showed that the real world, far from being the predictable universe imagined by others before him, is governed by chaos and could take a radically different path due to the smallest of changes. His work prompted the realization that apparently random or complicated behavior, found not only in meteorology but in many branches of natural science ranging from astronomy to ecology, do not necessarily require random or complicated underlying equations.

*Adam A. Scaife*

# TORNADOES

## the 30-second meteorology

### 3-SECOND BREEZE
A tornado is a whirling funnel of air extending from the base of a large cumulonimbus cloud, generating winds stronger than any other weather phenomenon.

### 3-MINUTE SHOWER
The USA is famous for its destructive tornadoes, giving an annual average bill of over $1 billion and, in 2011, killing 553. Other parts of the world are affected too—in 1989 Bangladesh suffered around 1,300 fatalities from a single tornado. Surprisingly, the UK and Holland are the most tornado-hit countries measured by frequency per square kilometer, though their tornadoes are generally much weaker than their U.S. counterparts.

**Strong, dry, westerly flow aloft** from the Rockies overlying a moist, warm, southerly feed from the Gulf of Mexico creates exactly the right unstable conditions over a swathe of central USA to generate long-lived, self-sustaining cumulonimbus clouds. Increasing horizontal winds with height separate the warm updraft from the cold, precipitation-induced downdraft, which hits the ground and scoops up warm, surface air into the updraft. Variation of horizontal wind with height causes rotation about a horizontal axis, like a pencil rolled between the hands. This horizontal vortex then tilts upright if it is drawn into the updraft and starts to produce rotation about a vertical axis. The resulting supercell storm continues sucking up warm, moist, low-level air to feed its insatiable, rotating updraft. Convergence of air from miles around concentrates and increases rotation exponentially with time in the way iceskaters spin ever faster by drawing their arms in. The pressure drop caused by high wind speeds and rapid ascent cools the air resulting in condensation, revealing the funnel of rotating air. If it touches down it becomes a tornado, marked by soil and debris lifted, whirled, and flung outward on its passage. The most damaging tornadoes—which can lift trucks and raze buildings—have winds exceeding 250 mph.

### RELATED TOPICS
See also
CLOUDS
page 22

HURRICANES & TYPHOONS
page 146

### 3-SECOND BIOGRAPHIES
TETSUYA THEODORE FUJITA
1920–98
Japanese-American meteorologist, who devised the Fujita scale which links tornado damage with wind speed

KEITH BROWNING
1938–
English meteorologist who coined the word supercell after studying a massive storm that affected the town of Wokingham in Berkshire, UK

### 30-SECOND TEXT
Edward Carroll

*Spin imparted to a column of air by wind shear, like a pencil rotated between the hands, is tilted vertical and concentrated by the persistent, powerful updraft in a supercell thunderstorm.*

# SUDDEN STRATOSPHERIC WARMING

## the 30-second meteorology

### 3-SECOND BREEZE
Every couple of years, gigantic breaking waves high in the atmosphere reverse the usual westerly winds and cause dramatic warming of the winter stratosphere.

### 3-MINUTE SHOWER
Sudden Stratospheric Warmings can also herald big changes at the Earth's surface: the eastern USA and Europe are often exposed to severe winter weather for weeks after an event. One occurred in the extreme winter from December 2009 to February 2010, with multiple impacts on society, ranging from transport disruption to increased energy demand across northern Europe. This seemingly obscure phenomenon is now an important clue in the development of long-range weather forecasts.

## By the mid-twentieth century

regular weather balloons were being launched from locations all over the world. Some balloons ascended 18 miles before bursting, providing measurements from well into the stratosphere. As often happens in science, these new observations yielded an entirely unexpected result. In January 1952 the temperature high over the Arctic suddenly increased by around 90°F in just a few days! This dramatic event, first reported by German researchers, is now, appropriately, named a Sudden Stratospheric Warming. Decades of observations since then show that Sudden Stratospheric Warmings occur every couple of years, but only in winter and almost exclusively over the Arctic. In 2002, a single surprise event over Antarctica temporarily filled the ozone hole. Gigantic, planetary-scale waves breaking in the stratosphere are responsible for these events, much like waves breaking on the beach (the Quasi-Biennial Oscillation results from a similar process involving small-scale waves). This disruption causes the normal west–east flow of winds around the Arctic to completely reverse, resulting in air falling in toward the North Pole where it is compressed. This compression, rather than any actual heating, causes the temperature to rise so dramatically.

### RELATED TOPICS
See also
LAYERS OF THE ATMOSPHERE
page 18

ATMOSPHERIC WAVES
page 54

POLAR VORTEX
page 68

### 3-SECOND BIOGRAPHIES
RICHARD SCHERHAG
1907–70
German meteorologist who first discovered the "explosive warming of the stratosphere" in 1952

TAROH MATSUNO
1934–
Japanese meteorologist and the first to explain how Sudden Stratospheric Warmings work

### 30-SECOND TEXT
Adam A. Scaife

*Sudden Stratospheric Warmings temporarily destroy the cold polar vortex at high altitude and increase the risk of extreme cold snaps at the surface.*

# RESOURCES

## BOOKS

*Atmosphere, Weather and Climate*
R. G. Barry and R. J. Chorley
(Methuen, 1968)

*Atmospheric Science: An Introductory Survey*
John M. Wallace and Peter V. Hobbs
(Academic Press, 2006, 2nd edn)

*Climate, History and the Modern World*
H. H. Lamb
(Routledge, 2nd edn 1995)

*Color and Light in Nature*
D. K. Lynch and W. Livingston
(Cambridge University Press, 2nd edn 2001)

*Fluid Dynamics of the Mid-Latitude Atmosphere*
Brian J. Hoskins and Ian N. James
(Wiley-Blackwell, 2014)

*Global Warming: the Complete Briefing*
John T. Houghton
(Cambridge University Press, 5th edn 2015)

*Large-Scale Disasters: prediction, mitigation and control* (See chapter by J. Pudykiewicz and G. Brunet, "The first hundred years of numerical weather prediction")
Mohamed Gad-El-Hak, ed.
(Cambridge University Press, 2008)

*Light and Colour in the Open Air*
Marcel G. J. Minnaert
(Eng. trans. Dover Publications, 1954)

*Measuring the Natural Environment*
Ian Strangeways
(Cambridge University Press, 2003, 2nd edn)

*Our Affair with El Niño: How We Transformed an Enchanting Peruvian Current into a Global Climate Hazard*
S. George Philander
(Princeton University Press, 2004)

*Prophet or Professor? The Life and Work of Lewis Fry Richardson*
Oliver M. Ashford
(Adam Hilger, 1985)

*The Role of the Sun in Climate Change*
D. V. Hoyt and K. H. Schatten
(Oxford University Press, 1997)

*Seamless Prediction of the Earth System: from Minutes to Months*
G. Brunet, S. Jones and P. M. Ruti, eds
(World Meteorological Organization, 2015)

*The Sun's Influence on Climate*
J. D. Haigh and P. Cargill
(Princeton University Press, 2015)

*The Thinking Person's Guide to Climate Change*
Robert Henson
(American Meteorological Society, 2014)

## PERIODICALS & MONOGRAPHS

Brunet, G., et al.: "Toward a seamless process for the prediction of weather and climate: the advancement of sub-seasonal to seasonal prediction," *Bull. Amer. Meteorol. Soc.* (2010), 91, 1397–1406

Henley, B. J., et al.: "A Tripole Index for the Interdecadal Pacific Oscillation," *Climate Dynamics* (2015): 10.1007/s00382-015-2525-1

IPCC (Intergovernmental Panel on Climate Change) *Climate Change 2013: The Physical Science Basis. Contribution of Working Group I to the Fifth Assessment Report of the Intergovernmental Panel on Climate Change* (Cambridge University Press, 2013) 1535 pp.

Knight, J. R., et al.: "A signature of persistent natural thermohaline circulation cycles in observed climate," *Geophysical Research Letters* (2005), 32, L20708

Lorenz, Edward, N.: "Deterministic Nonperiodic Flow," *J. Atmos. Sci.* (1963), 20, 130–141 (doi: http://dx.doi.org/10.1175/1520-0469(1963)020<0130:DNF>2.0.CO;2)

Nobre, Carlos, et al.: "Addressing the complexity of the Earth system," *Bull. Amer. Meteorol. Soc.* (2010), 91, 1389–1396

Scaife A. A., et al.: "Skilful long range prediction of European and North American winters," *Geophysical Research Letters* (2014), DOI: 10.1002/2014GL059637

Shapiro, Melvyn A., et al.: "An Earth-system prediction initiative for the 21st century," *Bull. Amer. Meteorol. Soc.* (2010), 91, 1377–1388

Waugh, D. W. and Polvani, L. M.: "Stratospheric polar vortices," in *The Stratosphere: Dynamics, Chemistry, and Transport*, Geophys. Monogr. Ser., 190, 43–57 (AGU, Washington DC, 2010)

## WEB SITES & ONLINE READING

www.atoptics.co.uk
For a host of images of atmospheric optics

www.metoffice.gov.uk
The UK's national weather service providing weather and climate change forecasts for the UK and worldwide

www.nasa.gov
National Aeronautics and Space Administration website for the latest news, images, and videos from the USA's space agency

www.noaa.gov
National Oceanic and Atmospheric Administration, a scientific agency for the USA that focuses on conditions in the oceans and the atmosphere

www.swpc.noaa.gov
NOAA's laboratory and service center, the Space Weather Prediction Center, monitors and forecasts space weather and provides space weather alerts for the USA

www.sciencedirect.com/science
Includes an excellent series of articles on monsoons and rainy seasons from Elsevier's *Encyclopedia of Atmospheric Sciences* (paywalled)

www.sciencemag.org/content/194/4270/1121
The paper that "proved" Milankovitch right, published in *Science* (1976) 194: 4270, 1121–1132, can be found via the above link (paywalled)

## NOTES ON CONTRIBUTORS

### EDITOR

**Adam A. Scaife** is head of Monthly to Decadal Prediction at the UK Met Office and honorary visiting Professor at the University of Exeter. He investigates mechanisms and predictability of weather and climate and has over 20 years experience in modeling the atmosphere with computer models. He has published around 100 scientific papers in leading journals and his recent studies include exciting new evidence for long-range predictability of winter weather. Adam was recently awarded the Lloyd's of London Science of Risk Research Prize for Climate Change research and the L. G. Groves Prize for Meteorology. He regularly communicates the latest meteorological science to the public via television, newspapers, and other media.

### FOREWORD

**Professor Dame Julia Slingo DBE FRS** is the UK Met Office Chief Scientist. Former posts during her long-term career in climate modeling and research include Director of Climate Research in NERC's National Centre for Atmospheric Science, at the University of Reading, where she remains a Professor of Meteorology. She has also worked at NCAR (National Center for Atmospheric Research) in Boulder, Colorado. In 2006 she founded the Walker Institute for Climate System Research at Reading, aimed at addressing the cross-disciplinary challenges of climate change and its impacts.

### CONTRIBUTORS

**Gilbert Brunet** obtained his Ph.D. in meteorology at Canada's McGill University in 1989. He is head of the Meteorological Research Division, Environment Canada, and former chair of the World Weather Research Programme at the World Meteorological Organization in Geneva (2007–14). He has been recognized as an expert in dynamical meteorology since his post-doctoral work in the Department of Applied Mathematics and Theoretical Physics at Cambridge University, UK, and the Laboratoire de Météorologie Dynamique at the École Normale Supérieure, Paris, France.

**Edward Carroll** has worked for the UK Met Office for more than three decades, initially spending six years as a weather observer. Since then he has gained a masters degree in weather, climate, and modeling, worked as a forecaster, as a lecturer at the Met Office College, and as a developer of forecasting applications. For the past 15 years Edward has been a shift Chief Forecaster.

**Leon Clifford** has a long-standing interest in climate science and meteorology. He graduated in physics-with-astrophysics before moving on to postgraduate research, studying the seas and the polar ice caps and their role in the climate using satellite remote sensing. He worked for many years as a journalist covering science, technology, and business issues. He edits a climate science website, reportingclimatescience.com.

**Chris K. Folland** is a UK Met Office Hadley Centre Science Research Fellow, Honorary Professor at the University of East Anglia, Guest Professor of Climatology at the University of Gothenburg, Sweden, and Adjunct Professor at the University of Southern Queensland, Australia. For 25 years he led research teams studying climate change and variability using observations and climate models, monthly to interannual forecasting, and climate data set development. He has had several awards and fellowships and was four times a Lead Author of the Intergovernmental Panel on Climate Change, sharing the Nobel Peace Prize in 2007.

**Dargan M. W. Frierson** is an Associate Professor in the Department of Atmospheric Sciences at the University of Washington in Seattle. He attended North Carolina State University and Princeton University. In his research, he discovered why the amount of tropical rainfall is greater in the northern hemisphere, and has uncovered new ways in which the climates in different parts of the planet are linked.

**Joanna D. Haigh CBE FRS** is Professor of Atmospheric Physics and co-Director of the Grantham Institute (Climate Change and the Environment) at Imperial College London. She has been fascinated by weather since childhood and has been lucky enough to follow a career in meteorology. Her particular expertise is in how solar and heat radiation interact with the atmosphere and in the physics of climate change.

**Sir Brian Hoskins CBE FRS** became the first Director of the Grantham Institute at Imperial College London in 2008, and now divides his time between it and the University of Reading, where he is Professor of Meteorology. He has degrees in mathematics from the University of Cambridge and spent post-doctoral years in the USA. His research is in weather and climate, in particular the understanding of atmospheric motion from frontal to planetary scales. Brian is a member of the scientific academies of the UK, USA, and China.

**Jeff Knight** is a senior climate research scientist at the UK Met Office Hadley Centre. He leads a group of scientists that studies climate variability and its representation in climate models. His interests include modes of global climate variability, such as the Atlantic Multidecadal Oscillation, atmospheric variability in the North Atlantic-European region, and seasonal and decadal prediction. His work has been recognized by the World Meteorological Organization and he contributed to the Fifth Assessment Report of the Intergovernmental Panel on Climate Change in 2014.

**Geoffrey K. Vallis** is Professor of Mathematics at the University of Exeter, prior to which he taught at Princeton University for many years. In his research he combines theoretical and numerical approaches to study fundamental problems in the circulation of the atmosphere and ocean. He is the recipient of the Adrian Gill Prize of the Royal Meteorological Society and a Wolfson Research Award from the Royal Society.

# INDEX

# ACKNOWLEDGMENTS

PICTURE CREDITS
The publisher would like to thank Shutterstock
for providing the majority of images used within
the illustrations, and the following people and
organizations. Every effort has been made to
acknowledge the pictures; however, we apologize
if there are any unintentional omissions.

Alamy/Lebrecht Music and Arts Photo Library: 23BL.
Courtesy Oliver Ashford: 34.
AWI—Alfred Wegener Institute for Polar and Marine
Research: 49B (map).
Edward Carroll: 27B.
European Southern Observatory: 97 (background).
fromoldbooks.org: 37TR.
Geographicus: 37C.
Dr Benjamin Henley, University of Melbourne,
Australia: 131T, 131B.
Courtesy Jericho Historical Society: 29CL.
Library of Congress, Washington DC: 56, 97BR,
139BR.
NASA: 51B (both), 87T, 147B.
NGA Images: 53B.
NOAA: 11, 31TR, 75TL, 76T, 93BL, 93C, 97TR,
97CL, 97CR, 99BR, 151C, 153BR, 160.
Sailko: 37B.
Science Photo Library: 136; Emilio Segre Visual
Archives/American Institute of Physics: 148.
Smabs Sputzer: 31TL

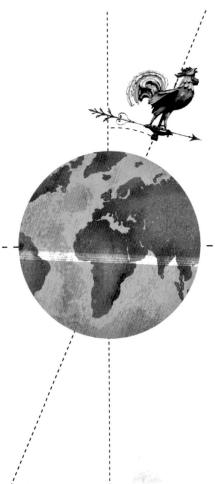